Low-Rate Wireless Personal Area Networks

...Enabling Wireless Sensors with IEEE 802.15.4™, Second Edition

José A. Gutiérrez

Edgar H. Callaway, Jr.

Raymond L. Barrett, Jr.

Published by
Standards Information Network
IEEE Press

Trademarks and Disclaimers

IEEE believes the information in this publication is accurate as of its publication date; such information is subject to change without notice. IEEE is not responsible for any inadvertent errors.

Library of Congress Cataloging-in-Publication Data

Gutiérrez, José A., 1967-

Low-rate wireless personal area networks: ...enabling wireless sensors with IEEE 802.15.4
/ José A. Gutiérrez, Edgar H. Callaway, Jr., Raymond L. Barrett, Jr. -- 2nd ed.
   p. cm.
Includes bibliographical references and index.
ISBN 0-7381-4977-2
1. Wireless LANs. 2. Sensor networks. I. Callaway, Edgar H. II. Barrett, Raymond L.,
1944- III. Title.

TK5105.78.G88 2007

004.6'8--dc22

2006049654

IEEE
3 Park Avenue, New York, NY 10016-5997, USA

Don Messina, Program Manager, Document Development
Jennifer McClain, Managing Editor
Linda Sibilia, Cover Designer

Review Policy

The information contained in IEEE Press/Standards Information Network publications is reviewed and evaluated by peer reviewers of relevant IEEE Technical Societies, Standards Committees and/or Working Groups, and/or relevant technical organizations. The authors addressed all of the reviewers' comments to the satisfaction of both the IEEE Standards Information Network and those who served as peer reviewers for this document.

The quality of the presentation of information contained in this publication reflects not only the obvious efforts of the authors, but also the work of these peer reviewers. The IEEE Press acknowledges with appreciation their dedication and contribution of time and effort on behalf of the IEEE.

To order IEEE Press Publications, call 1-800-678-IEEE.

Print: ISBN 0-7381-4977-2   SP1150

See other IEEE standards and standards-related product listings at: http://standards.ieee.org/

Trademarks

Bluetooth is a registered trademark of Bluetooth SIG, Inc. (www.bluetooth.org/).

DeviceNet is a trademark of ODVA (www.odva.org/).

Fieldbus is a registered trademark of the Fieldbus Foundation (www.fieldbus.org/).

HART is a registered trademark of the HART Communication Foundation (http://www.hartcomm.org/).

IEEE and 802 are registered trademarks of the Institute of Electrical and Electronics Engineers, Incorporated (www.ieee.org/).

IEEE Standards designations are trademarks of the Institute of Electrical and Electronics Engineers, Incorporated (www.ieee.org/).

INCOM is a trademark of Eaton Corporation (www.eaton.com/).

IrDA is a registered trademark of the Infrared Data Association (www.irda.org/).

ZigBee Alliance is a trademark of Philips Corporation (www.zigbee.org/).

Dedication

To my newborn son Alan Alejandro and my wife Natacha, for bringing inspiration at every instant of my life.

José A. Gutiérrez

To my parents, Pat and Ed, Sr., for the example they set and the guidance they gave; and to my wife, Jan. Without her support and understanding, this book would not have been possible.

Ed Callaway

To my wife, Julie, and our two wonderful children, Gloria and Raymond, who bring the meaning to my life and works.

Ray Barrett

Acknowledgment

Behind this book is the work of the larger IEEE 802® community, in particular, the outstanding work performed by the technical editorial team of the IEEE 802.15.4™ Task Group, led by the indomitable and indefatigable Dr. Bob Heile.

We would like to express our sincere thanks to a diverse group of persons who provided their support during the standardization process, and with whom we have had many enriching technical discussions: Anthony Allen, Venkat Bahl, Daniel Bailey, Phil Beecher, Monique Bourgeois (IEEE Std 802.15.4 MAC Technical Editor and IEEE Std 802.15.4b Technical Editor), Priscilla Chen, Francois Chin, Dr. Neiyer Correal, Robert Cragie, David Cypher, Ralph D'Souza, Dr. Larry Dworsky, James Gilb, Eric Gnoske (IEEE Std 802.15.4b Secretary), Paul Gorday, David Hayes, Dr. Lance Hester, Dr. Ivan Howitt, Dr. Jian Huang, Dr. Yan Huang, Øyvind Janbu, Phil Jamieson (IEEE Std 802.15.4 Vice Chair), Pat Kinney (IEEE Std 802.15.4 Chair), Masahiro Maeda, Dr. Fred Martin, Dave McClanahan, Vinay Mitter, Said Moridi (IEEE Std 802.15.4 PHY Technical Editor), Dr. Bob O'Dea, Luis Pereira, Dr. Robert Poor (IEEE Std 802.15.4b Chair), Clinton Powell (IEEE Std 802.15.4b PHY Technical Editor), Larry Puhl, Joseph Reddy, Phil Rudland, Dr. Chip Shanley, Qicai Shi, Zachary Smith, Bob Stengel, Rene Struik, David Taubenheim, Hans van Leeuwen, and Dr. Andreas Wolf. In addition to many of the above, Farron Dacus, Barbara Doutre, Latonia Gordon, Charles Luebke, and Marco Naeve (IEEE Std 802.15.4 Secretary and IEEE Std 802.15.4b Vice Chair) commented on the many early drafts of this book.

We also would like to thank the anonymous peer reviewers of this book for their evaluation and guidance in helping to enrich the material presented here.

Finally, we would like to thank Jennifer Longman, Angela Ortiz, and the rest of the IEEE editorial team for their diligence and the great work performed.

Authors

José A. Gutiérrez is Corporate Director of Technology at Emerson Electric, leading a wide range of technology activities that include the use of wireless sensor networks in the industrial environment. He received the B.S. degree in electronic engineering from Universidad Simón Bolivar in Caracas, Venezuela, in 1991, M.S. in electrical engineering from the University of Milwaukee - Wisconsin in 2001, and the Ph.D. degree from the same university in 2005. Through his active membership at the IEEE LAN/MAN Standards committee, in 2000 Dr. Gutiérrez became the Editor in-Chief of the
IEEE 802.15 Working Group, Task Group 4, focused in the development of Low-Rate Wireless Personal Area Networks. After the success of the 802.15.4 standard he supported the creation of the ZigBee Alliance where from 2003 to early 2005 he worked as program manager. Currently he is a member of the board of directors of the Wireless Industrial Network Alliance and active member of the ISA SP100 committee developing a new standard for Industrial Wireless Networks. Dr. Gutiérrez has multiple patents in the area of wireless personal and local area networks and author of multiplicity of papers in the areas of automatic control, artificial intelligence, and wireless communications. Currently he is working on his next book focused on Robotic Imaging.

Edgar H. Callaway, Jr. received the B.S. degree in mathematics and the M.S.E.E. degree from the University of Florida in 1979 and 1983, respectively, the M.B.A. degree from Nova (now Nova-Southeastern) University in 1987, and the Ph.D. degree in computer engineering from Florida Atlantic University in 2002. He joined the Land Mobile Division of Motorola in 1984 as an RF engineer working on trunked radio products and in 1990 transferred to Motorola's Paging Products Group, where initially he designed paging receivers for the Japanese market. From 1992 to 2000 he was engaged in paging receiver and transceiver system design, and was the lead receiver designer of Motorola's paging platform. In 2000, he joined Motorola Labs, where his interests include the design of low-power wireless networks. He is a Registered Professional Engineer (Florida), and has more than thirty issued U.S. patents. Dr. Callaway is the author of *Wireless Sensor Networks: Architectures and Protocols* (Auerbach, 2004) and several papers and book chapters.

Raymond Louis Barrett, Jr. received the B.S.E.E. degree from Case Institute of Technology (now Case-Western University) in 1966, the M.S. in computer science and M.B.A. degrees from Nova (now Nova-Southeastern) University in 1983 and 1982, respectively, and the Ph.D. in electrical engineering from Florida Atlantic University (FAU) in 1990. He taught as an Assistant Professor for Nova University and FAU, and as an Associate Professor at the University of North Florida from 1982 through 2001. Concurrently with teaching at FAU, he joined  the Motorola's Paging Products Group in 1991 as a Senior Member of the Technical Staff whereupon he initially designed various low-power paging receiver circuits. In 1999 he joined Motorola Labs, where his interests included the development of low-power wireless components. He is presently a Principal Member of the Technical Staff with MAXIM Integrated Products working on portable products. He is a Registered Professional Engineer (Florida) and has more than thirty issued U.S. patents and several published papers.

Contributor

Marco Naeve is a senior engineer for the Embedded Systems and Communications Group of the corporate research and development unit (Innovation Center) at Eaton Corporation. He received his B.S. degree in electrical engineering from the Milwaukee School of Engineering, Wisconsin in 1996. He also holds the equivalent degree of a Diplom Ingenieur (FH) in electrical engineering with focus on RF and telecommunication systems, which he received from the University of Applied Science Lübeck in Germany. His work at Eaton Corporation includes modeling and simulation  of low-rate wireless personal area networks, wireless local area networks, and wired industrial communication systems. He is an active member of the IEEE and currently holds the position of vice-chair of the IEEE 802.15.4b LR-WPAN task group and was actively involved in the developments of the IEEE 802.15.4-2003 and IEEE 802.15.4-2006 standards. He was also a member of the editing team for ZigBee's network layer specifications and is currently the vice-chair of the Zig-Bee application framework group. Marco co-authored two papers and has three patent applications pending.

Foreword

There is a program offered by the IEEE Standards Association called IEEE Get 802®. Through this program, IEEE 802 standards are available to the public at no cost six months after publication. Within two years of its initial publication in 2003, IEEE Standard 802.15.4™ became the third most downloaded 802 standard behind IEEE Std 802.11™ and IEEE Std 802.3™. On the ZigBee™ Alliance web site you can download the ZigBee specification, which is built on 802.15.4. Downloads have been averaging a steady 500 per week. Clearly interest in 802.15.4 is intense. I was delighted when the first addition of *Low-Rate Wireless Personal Area Networks* hit the market. Having a concise, clearly written guide is invaluable to helping developers understand a lengthy standards document. Just as important is helping the market understand the application space and practical issues surrounding real world deployment. This book does it all, and the second edition brings us up to date on a rapidly evolving and growing market.

The authors of this book are uniquely qualified for the task because not only are they intelligent, witty, and articulate, they are inextricably bound to the creation of the standard in the first place. IEEE 802.15 as a Working Group was started with the initial thought of developing a standard for a low data rate class of radio. Around the same time, the Bluetooth® SIG was announced and for a while the marketplace was confused as to whether Bluetooth would address the needs of sensor and control networks. By late 2000, it was clear that while Bluetooth was very effective for wireless headsets, it was not the solution for sensor networks which needed much lower cost, very long battery life, and support for mesh networking. The continuing need for a standard in this space led to the formation of a Task Group in 802.15 to draft what became IEEE Std 802.15.4.

I had the honor of chairing the initial activity, where José Gutiérrez learned a valuable lesson on how one gets volunteered in spite of oneself, and he became the group's technical editor. This gives him a unique perspective on every aspect of the standard. Additionally, Ed Callaway, Ray Barrett, Venhat Bahl, and Kursat Kimyacioglu all had significant background in low data rate radios and mesh networking. While there were many contributors, these gentlemen in particular worked tirelessly to merge this understanding into what became the technology baseline for 802.15.4. I can not imagine a better threesome than José, Ed, and Ray to create a useful guide and keep it up to date. In addition, Ed and José both contributed to the formation of the ZigBee Alliance. José served as the head of the Alliance Program Management Office, and Ed was on the Alliance Board of Directors. All of them remain active to this day in various aspects related to the deployment and enhancement of the technology.

Since the first addition of *Low-Rate Wireless Personal Area Networks*, much has happened. We launched a Physical Layer amendment called 802.15.4a™, which is in the final stages of balloting in the IEEE Standards process. We also completed a revision to the standard, which was published this year as IEEE Std 802.15.4™-2006. This revision cleaned up some issues in the MAC, substantially improved the security suite, and added higher data rate modulation modes to the sub-1 GHz bands. We are also about to start two new amendments, 802.15.4c™ and 802.15.4d™, to add sub-1 GHz band coverage in China and Japan, in addition to those currently supported in the USA and Europe. Also since the first addition, the ZigBee Alliance has published its specification, which adds the networking stack and application layer to 802.15.4. The ZigBee spec had been downloaded over 35,000 times as of October 2006. The bottom line of all this activity is that 802.15.4 is likely to become the most widely deployed radio on the planet over the next decade, because it will wind up in so many things.

Whether you are trying to get a first pass feel of what the market is all about or whether you are trying to get a better intuitive grasp of the standard and how it works, this is the place to come. Thanks guys, and if I do my job right, you better start making plans for the third edition.

Bob Heile
Chair, IEEE 802.15
Chairman, ZigBee Alliance

Table of Contents

Introduction

Since the publication of the IEEE 802.15.4 standard in 2003, the interest in low-rate wireless personal area networks and wireless sensor networks has increased significantly. The market saw a large number of new companies emerge, providing products and solutions while original equipment manufacturer (OEMs) are looking for new opportunities for growth in this emerging space. Hardware that employs the ZigBee protocol stack implemented on the IEEE Std 802.15.4 PHY and MAC layers can be found in environmental controls, heart patient monitors, and commercial lighting controls. While more and more products appear on the market, the IEEE Std 802.15.4 continues to be a work in progress. Following the initial release in 2003, two new IEEE 802.15 task groups emerged to continue the work in this area. The first is IEEE 802.15.4a, which is working towards an amendment of the standard with the goal to provide an alternate physical layer with additional capabilities such as precision ranging and location as well as high aggregate throughput and ultra low power capabilities. The result of the second, the IEEE 802.15 task group 4b is a revision of the original standard which was published in 2006. As part of this revision, the PHY layer definition has been expanded to include new modulation schemes. Higher data rates are now added in some bands, and more channels have been added. The new capabilities are in addition to the existing definitions and ensure compatibility with existing systems. The MAC layer definition has been augmented to include new components that support the new PHY modulation schemes. Much of the modification to IEEE Std 802.15.4 is occurring in response to the rapid market acceptance of products using this and competing technologies in the same, or overlapping spectra.

This book is designed to complement the IEEE 802.15.4 standard by presenting an overview of the features that characterize it, the applications that motivate it, and the rationale that dictated several of the design options chosen during its creation. The standard is written and structured in a way that makes it very easy to understand and (we believe) easy to implement. The IEEE 802.15.4 technical editing team concentrated especially on creating informative clauses to improve the communication of the more complex concepts found in the standard. This book does not pretend to extend that effort except in discussion of applications, but rather intends to be a companion work. It is directed to people interested in the field of "simple" wireless connectivity. It has a focus in wireless sensors and actuators for the industry in general.

The material in this book is divided into three parts. Part I presents an overview of the low-rate wireless personal area technology and IEEE Std 802.15.4. The material focuses on the motivation for the creation of this standard, including the application scenarios that drove the making of it. Part I is designed not only as a technical introduction to the standard, but also to offer marketing and business professionals enough background information on the technology and the vision behind its conception to help plan marketing and business strategies.

Part II concentrates on the technical features and components of the IEEE 802.15.4 standard. It adds material on network layer functionality that is not covered by the standard. The network layer information applies to the rationale behind the design of several of the components of the standard for the envisioned applications. Part III focuses on implementation and system design considerations. It includes an analysis of system-level real-world issues that will be important for prospective implementers to consider.

The information in IEEE Std 802.15.4 is compact and simplicity is reflected in the text. The standard is organized into seven clauses and seven annexes. Clauses 1 though 5 contain introductory material that maps with Part I of this book. Clauses 6 and 7 in the standard define the physical layer and medium access control (MAC) sublayer, respectively; these clauses map with Part II. Finally, as explained above, Part III does not map to the standard, since it deals with implementation and real-world issues that the standard does not address.

Acronyms and Abbreviations

ACK	acknowledgment
ACL	access control list
AES	advanced encryption standard
ASK	amplitude shift-keying
BPSK	binary phase shift-keying
BSP	beacon synchronization parameter
CAP	contention access period
CBC-MAC	cipher block chaining message authentication code
CCA	clear channel assessment
CCM	counter mode + CBC-MAC
CFP	contention-free period
CRC	cyclic redundancy check
CSMA-CA	carrier sense multiple access with collision avoidance
CTR	counter mode
DSSS	direct sequence spread spectrum
ED	energy detection
ETSI	European Telecommunications Standards Institute
EVM	error-vector magnitude
FCC	Federal Communications Commission
FCS	frame check sequence
FFD	full function device
FHSS	frequency hopping spread spectrum
GSM	global system for mobile communication

GTS	guaranteed time slot
HVAC	heating, ventilation, and air conditioning
IC	integrated circuit
IrDA	infrared data association
ISM	industrial, scientific, and medical
ITU-T	International Telecommunication Union - Telecommunication Services
LAN	local area network
LC	inductor-capacitor
LLC	logical link control
LQI	link quality indication
LR-WPAN	low-rate wireless personal area network
LSB	least significant bit
MAC	medium access control
MCPS-SAP	MAC common part sublayer service access point
MFR	MAC footer
MHR	MAC header
MIC	message integrity code
MIPS	million instructions per second
MLME	MAC sublayer management entity
MLME-SAP	MAC sublayer management entity service access point
MSB	most significant bit
MPDU	MAC protocol data unit
MSDU	MAC service data unit
MSK	minimum shift-keying
OEM	original equipment manufacturer
O-QPSK	offset quadrature phase shift-keying

OSI	open systems interconnection
PAN	personal area network
PD-SAP	PHY layer data service access point
PHY	physical
PHR	PHY header
PIB	PAN information base
PLME	PHY layer management entity
PLME-SAP	PHY layer management entity service access point
PN	pseudo-noise
POS	personal operating space
PPDU	PHY protocol data unit
ppm	parts per million
PSDU	PHY service data unit
PSSS	parallel sequence spread spectrum
QoS	quality of service
RAM	random access memory
RC	resistor-capacitor
RF	radio frequency
RFD	reduced function device
RFID	radio frequency identification
RSSI	received signal strength indication
RX	receive or receiver
SAP	service access point
SAW	surface acoustic wave
SHR	synchronization header
SNR	signal-to-noise ratio

SoC	system-on-a-chip
SSCS	service specific convergence sublayer
TDMA	time division multiple access
TX	transmit or transmitter
VCO	voltage controlled oscillator
WLAN	wireless local area network
WPAN	wireless personal area network
WSN	wireless sensor network

Part I

Chapter 1   The IEEE Standard 802.15.4

...enabling wireless sensor networks

The explosive growth of embedded control and monitoring in almost any electronic device and the need for connectivity of these applications is causing an integration bottleneck. Manufacturers use different communication interfaces—standard or proprietary—creating the need for yet another controller task to make the arbitration. Conventionally, these communication links are wired. Wires allow power and the reliable transmission of signals from a controller to its peripherals. When the peripherals are not physically contained in the controller, the required wiring brings issues such as cost of installation, safety, and operation convenience to the surface. Wireless technology is the obvious solution to overcome these obstacles, although it comes with its own set of challenges—propagation, interference, security, regulations, and others. The technology to overcome these issues exists, but normally with added complexity causing an increase in the cost of the system.

Certainly some applications can afford the cost of adding a high-end wireless communication system such as cellular phones, IEEE Std 802.11™ Wireless Local Area Networks, and so on. Conversely, many other applications can be enhanced or enabled if a standard low-cost wireless communications solution was available. A standard would drive interoperability among different manufacturers with direct benefit to the end consumer.

A low-rate wireless personal area network (LR-WPAN) is a network designed for low-cost and very low-power short-range wireless communications. This definition is at odds with the current trend in wireless technologies whose focus has been on communications with higher data throughput and enhanced quality of service (QoS).

The current trend in wireless technologies is to produce protocols with higher data throughput and QoS as their primary performance metrics, with cost and power consumption secondary. The primary performance metrics of an LR-WPAN are cost and power consumption, with data throughput and QoS secondary.

WLANs, WPANs, and LR-WPANs

Wireless Local Area Networks (WLANs) were created as the wireless extension of the IEEE 802® wired local area network (LAN), which was designed for high-end data networking. Among the system requirements of a WLAN are seamless roaming, message forwarding, longest possible range and capacity for a large population of devices. In contrast, WPANs are designed to function in the Personal Operating Space (POS), extending up to 10 m (30 ft) in all directions and covering the area around a person whether stationary or in motion.

WPANs are used to convey information over relatively short distances among the participant transceivers. Unlike WLANs, connections effected via WPANs involve little or no infrastructure. This allows small, power efficient, inexpensive solutions to be implemented for a wide range of devices.

The IEEE 802.15 Working Group has defined three classes of WPANs that are differentiated by data rate, battery drain, and QoS. The high-data rate WPAN (IEEE Std 802.15.3™) is suitable for multimedia applications that require very high QoS. Medium-rate WPANs (IEEE Std 802.15.1™/Bluetooth®) are designed as cable replacements for consumer electronic devices centered on mobile phones and PDAs with a QoS suitable for voice applications. The last class of WPAN, LR-WPAN (IEEE Std 802.15.4™), is intended to serve applications enabled only by the low power and cost requirements not targeted in the previous WPANs. LR-WPAN applications have a relaxed need for data rate and QoS. Figure 1–1 illustrates the operating space of the 802 WLAN and WPAN standards. Notice that IEEE Std 802.15.4 is not designed to overlap with higher end wireless networking standards.

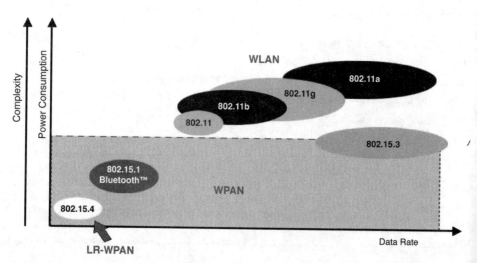

Figure 1–1: Operating space of various WLAN and WPAN standards

Table 1–1 presents a summary of the principal characteristics of an LR-WPAN using IEEE Std 802.15.4, compared with IEEE Std 802.11b™ and a standard WPAN such as IEEE Std 802.15.1.

The intent of IEEE Std 802.15.4 is not to compete with other wireless networking technologies but to complement the range of available wireless technologies in the lower end of the spectrum of data rates, power consumption, and cost.

Although possible for certain applications, IEEE Std 802.15.4 was not designed to overlap its application space with other wireless networking standards.

The authors strongly discourage the use of IEEE Std 802.15.4 for conventional WLAN applications. It may result in a highly challenging endeavor (with very long latencies).

Table 1–1: Comparison of LR-WPAN with other wireless technologies

	802.11b WLAN	Bluetooth™ WPAN	Low Rate WPAN
Range	~100 m	~10 - 100 m	10 m
Data Rate	~2-11 Mb/s	1 Mb/s	≤ 0.25 Mb/s
Power Consumption	Medium	Low	Ultra Low
Size	Larger	Smaller	Smallest
Cost/Complexity	Higher	Medium	Very Low

LR-WPAN technology pursues applications where WPAN solutions are still too expensive, extremely low-power operation is needed, and/or the performance of a technology such as Bluetooth is not required.

LR-WPANs complement the range of wireless networking technologies by providing very low-power consumption capabilities at very low implementation cost while enabling applications that were previously impractical or implemented with proprietary technologies.

ISO/OSI REFERENCE MODEL

IEEE 802 communication standards define only the bottom two layers of the International Standard Organization's (ISO) Open System Interconnection (OSI) protocol reference model [26]: the physical (PHY) layer and the data link layer [17]. The other layers are not specified in the standard and are normally specified by industrial consortia formed by companies interested in the manufacturing and use of the particular standard.

 The layered reference model allows the encapsulation of different levels of abstraction with a well-defined functionality. Each layer offers services to the layer above through the use of service primitives.

In the case of IEEE Std 802.15.4, the ZigBee™ Alliance is an organization that has led the development of the upper layers, through the definition of application profiles. These profiles make use of a simplified five layer ISO/OSI reference model as shown in Figure 1–2.

Seven Layer ISO-OSI Model		Simplified Five Layer ISO-OSI Model	IEEE 802 model
7	Application	User Application	Upper Layers
6	Presentation	Application Profile	
5	Session		
4	Transport		
3	Network	Network	
2	Data Link	Data Link	Logical Link Control (LLC)
			Media Access Control (MAC)
1	Physical	Physical	Physical signaling (PHY)

Figure 1–2:   ISO-OSI reference model and IEEE 802 standards model

WIRELESS SENSOR NETWORKS

Wireless Sensor Networks (WSNs) are a subset of wireless networking applications focused on enabling connectivity without the use of wires to sensors and actuators in general.

The ZigBee Alliance is a consortium formed by several leading semiconductor and industrial manufacturers, distributors, and end users.

One of the tasks of this organization is the definition of the networking support and applications profiles that will use IEEE Std 802.15.4-compliant transceivers.

Sensor networks can be classified according to the type of sensors, kind of application (industrial, medical, vehicular, etc.), environment where the network operates (explosive, vibration, acceleration, temperature, etc.), and by network parameters (network topology, required throughput, range, etc.). The IEEE 802.15.4 Working Group is chartered to focus on wireless sensor networks.

 Due to the length of the name "Wireless Sensor & Actuator Networks," the industry has adopted "Wireless Sensor Networks" instead. In any case, it is important to remember that the design of this type of network is meant to collect and to send information to wireless transceivers attached to a sensor and/or actuator.

The interest in WSNs has three faces. First, there is a need to lower the cost of sensor installation, which comes in the form of cabling, labor, materials, testing, and verification. For example, a limit-switch may cost less than a dollar, but the installation cost may range from $50 to $100. Furthermore, wiring regulations in the industrial and residential environment may require additional materials and activities for the wire installation, e.g., conduit and installation labor.

Second, cables require connectors that can get loose, lost, misconnected, or break (due to frequent access to neighboring devices). This problem is commonly known as "the last meter connectivity problem" and is named this way due to the analogous problem in the wide area networking called "the last mile connectivity problem."

The third aspect is that WSNs form the lower layer of intelligent maintenance systems enabling sensor-rich environments that generate abundant data that may be used to improve industrial operations. WSNs allow the gathering of more frequent data on a large number of machines and industrial systems in general. The use of large amounts of hardwired sensors networked to central systems brings a lot of complexity to the system, making it impractical in most cases.

A wireless solution for sensor networks provides flexible connectivity without the need for connectors. In addition, improved operator safety can be achieved by the use of wireless systems. Of course, WSNs share most of the issues surrounding wireless applications such as information security, authentication, small-scale radio-frequency propagation, antenna placement, and others.

Mobility is another benefit from the wireless solution, although in the WSN context, this capability is traded with "ease of installation." In other words, mobility is normally not a requirement for a WSN system, but certain mobility concepts can be used to enable ad hoc networking. It is important to clarify that the term mobility in this context refers to relative motion of devices with respect to each other. For example, a wireless network within a moving machine is not considered mobile if the sensors are meant to communicate only with devices within that machine.

The set of advantages described is not enough to replace hardwired connections. The reliability and security (perceived and real) of wired networks can be higher than wireless communication systems. It is expected, however, that hybrid networks, wired and wireless, will coexist. Wireless sensors will act as extensions of wired networks wherever the wireless capability adds value to the specific application.

Challenges in the Design of Wireless Sensor Networks

The inertia slowing the widespread implementation of WSNs is the lack of standardized technologies that can address their requirements both at the application level and from the communications point of view. The focus of the wireless industry has been primarily on communications with higher data throughput, leaving short-range wireless connectivity behind.

An important feature required for WSNs is the capacity for easy installation of a large number of transceivers. This requirement itself has all of the issues surrounding wireless communication (e.g., range), plus other challenges unique to the application. An example of these challenges is the need to logically bind specific devices together—a lamp and its light switch.

Among the possible wireless communication technologies, we have the following:

- *Light-communications:* This includes technologies like the Infrared Data Association (IrDA®) standard among others. The main disadvantage of this technology is the need for an unobstructed line-of-sight among the devices operating in this type of network.

- *Inductive Fields:* This technology has been extensively used for Radio Frequency Identification (RFID) applications. The main disadvantages of this technology are very low range and high energy required by the network coordinator. In addition, field alignment may be required for efficient communications.

- *Ultrasound:* Similarly to inductive fields, ultrasound requires high energy from the network coordinator. Line-of-sight is not a major concern in this technology but form factor (miniaturization) is.

- *Radio-frequency (RF):* This technology does not require an unobstructed line-of-sight. The current state of the technology allows the implementation of low-power radio transceivers with both data-rates and ranges scalable according to the application.

RF technologies seem to provide a sufficient set of advantages for the implementation of wireless sensor networks, but they also have a set of challenges that need to be addressed. Some of the high-level issues in the design and implementation of WSNs are described in the following section.

Power Consumption

Some applications require the use of completely untethered RF transceivers (no access to external power); this implies the use of batteries or power scavenging. If batteries are used, they should last a long time, because the need for sensor maintenance due to battery replacement goes against the original

> WSNs, in general, cannot make use of the "battery recharging" culture commonly associated with consumer electronic devices such as mobile phones and PDAs.

intention of ease of installation and low-cost operation. A direct implication of limited power is a limited communication range.

A common solution to overcome this obstacle is to use power cycling; that is lower the duty cycle of operation of the device (this solution brings to surface a synchronization problem, which can be solved with the appropriate networking procedures).

Some elementary calculations reveal that a AAA battery with capacity of 750 mAh powering off-the-shelf short-range radio transceivers (10 mA typical active current consumption) will last for two years if a duty cycle of less than 0.5% is maintained.

Range

Due to governmental regulations and implementation economics, RF power outputs ranging from 0 dBm to 20 dBm are typical of wireless systems operating in

> The design of IEEE Std 802.15.4 allows the implementation of radio transceivers with extremely low power consumption in comparison to current wireless networking radio technologies.

unlicensed bands. The limited power establishes a limited connectivity range—the maximum distance between a transmitter-receiver pair is constrained.

In the context of WSNs, multihop network protocols are required to circumvent this problem; this in turn implies the need for suitable routing algorithms.

 An RF output power of 0 dBm with radios having a sensitivity of −70 dBm will have a range of 10 m (30 ft.) for average indoor environments (using a log-distance path loss model with path loss exponent equal to 3).

Availability of Frequency Bands

The RF spectrum is a scarce resource regulated by most governments. However, there are special unlicensed bands, the use of which is allowed if the wireless devices operate within a set of rules that control the RF output envelope in time, frequency, and amplitude. In some bands, in some regulation regions, regulations stipulate that in order to transmit a larger RF output power, the energy may have to be spread—i.e., spread spectrum modulation should be employed. The use of unlicensed bands is free of charge, provided that the manufacturer can demonstrate conformance with the rules.

For WSNs, the most common bands used (or planned to be used) are:

- *868.0–868.6 MHz:* Available in most European countries
- *902–928 MHz:* Available in North America
- *2.40–2.48 GHz:* Available in most countries worldwide
- *5.7–5.89 GHz:* Available in most countries worldwide.

Radio devices operating in unlicensed bands must comply with local regulations. In the U.S., the Federal Communications Commission (FCC) is the institution in charge of the regulation of these devices. Similarly, the European Telecommunications Standards Institute (ETSI) coordinates regulation efforts in Europe. Other countries around the world have their own regulatory agencies; many of them accept either FCC or ETSI acceptance as proof of compliance.

IEEE Std 802.15.4 contains an informative annex focused on regulatory requirements worldwide. The material, although introductory in nature, provides the implementer with enough background information to initiate the certification endeavor.

Wireless networking standards operating in the global unlicensed bands offer potentially lower cost due to technology availability, high volumes, and limited or no reengineering (no re-tuning). Recently, the major focus of short-range wireless communication Original Equipment Manufacturers (OEMs) has been in the 2.4 GHz and 5.7 GHz bands due to the large bandwidth available.

The previous remarks identify a coexistence challenge because incompatible technologies sharing the same band will be competing to gain and maintain access. Several studies have been conducted to evaluate the impact of this issue, revealing potential problems that depend on traffic load, spatial distribution, output power, transceiver density, and RF propagation parameters [3], [9–12], and [27].

IEEE Std 802.15.4 was designed to operate in the following bands:

868.0 to 868.6 MHz → 1 Channel (20 kb/s, 100 kb/s, 250 kb/s)

902.0 to 928.0 MHz → 10 Channels (40 kb/s, 250 kb/s)

2.40 to 2.48 GHz → 16 Channels (250 kb/s)

The first two bands combined are called the "low band," and the third is called the "high band."

Network Topology

To overcome the challenge of limited range, multihop network topologies are required. A special requirement for WSNs is the need for low maintenance. To achieve this goal, the network topology should be designed to allow low duty cycle operation among the member devices of the network. An analysis of network topologies and network layers suitable for WSNs is presented in Chapter 6.

At this point, it is important to note an often overlooked fact: Wireless Networking implies bidirectional communications among the devices participating in the network. This characteristic in turn increases the reliability of the system and allows monitoring of the state of all the devices.

Self-Organization

To achieve ease of installation, it is required that the wireless network be self-organizing, that is, each sensor device start participating in a network without requiring special in-situ configuration (i.e., addressing, association, and traffic balancing).

Self-organization is a feature of ad hoc networks. Ad hoc wireless networks are a collection of transceivers that form a network without the aid of any fixed infrastructure or centralized administration. The network employs routing protocols to determine the proper message path from source to destination; for example, in one protocol, each radio keeps a list of adjacent radio links and updates a routing table when the network topology changes.

The implementation of a self-organization strategy is a mixture among network topology, security implementation and application requirements. This feature is commonly addressed by the definition of application profiles, which are not part of IEEE Std 802.15.4. Application profiles group together applications with common functionality to achieve interoperability among the devices participating in it.

Chapter 2   Low-Rate Wireless PAN Applications

...motivation for a vision

IEEE Std 802.15.4 was designed to be used in a wide variety of applications, requiring simple wireless communications over short-range distances with limited power and relaxed throughput needs. The applications that IEEE Std 802.15.4 addresses can be placed in the following classifications:

- *Stick-On Sensor:* These applications comprise wireless sensors that can operate completely untethered. This implies battery-powered transceivers. The focus of these applications is for monitoring or remote diagnostics.

- *Virtual Wire:* These are monitoring and control applications that can only be enabled through wireless connectivity, in places where a wired communications link cannot be implemented, e.g., tire pressure monitoring, motor bearing diagnosis, engine components, and so on.

- *Wireless Hub:* These are applications in which a centralized wireless bridge is added over a wired network. A wireless hub acts as a gateway between a wired network and a wireless LR-WPAN network. In many cases, the wireless hub network is formed with two transceivers, one being the hub and the other being an LR-WPAN device embedded in a device like a PDA.

- *Cable Replacement:* These applications attempt to add value through the removal of wires in consumer electronic portable devices. Some of these applications have been addressed by another WPAN technology: IEEE Std 802.15.1 (Bluetooth). In this case, IEEE Std 802.15.4 offers a low power alternative (at least one order of magnitude lower) with the other benefits of reduced cost. Cable replacement differs from "Stick-On" sensor applications in that the former may use a continuous source of power (mains) or rechargeable batteries.

Figure 2–1 shows the LR-WPAN application scenarios just introduced.

It is important to point out that the focus of IEEE Std 802.15.4 is to be an application enabler, where the value will be in the application and not in the wireless capability. This vision comes at odds with others, where the application is the communications capability—e.g., cellular telephony.

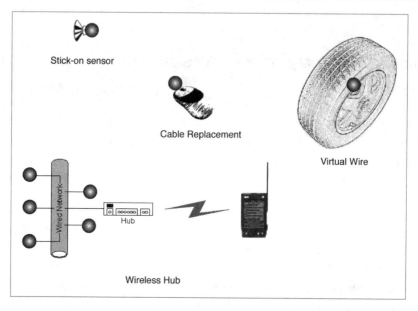

Stick-on sensor

Cable Replacement

Virtual Wire

Wired Network

Hub

Wireless Hub

Figure 2–1: Some LR-WPAN application scenarios

This chapter examines some of the IEEE Std 802.15.4 applications that several leading companies and visionaries have proposed and that have served as motivation for the creation of this standard.

 IEEE Std 802.11, IEEE Std 802.15.1, and IEEE Std 802.15.3 were created to target specific applications. In contrast, IEEE Std 802.15.4 was designed to address a wide range of applications in different market segments.

INDUSTRIAL AND COMMERCIAL CONTROL AND MONITORING— WIRELESS SENSORS

As previously introduced in Chapter 1, the interest in wireless connectivity for the industrial and commercial sector is centered on the need to lower the installation cost of sensors and actuators, while creating a sensing-rich environment as a lower layer for intelligent systems.

Initial implementations using IEEE Std 802.15.4 will most likely occur in monitoring applications with non-critical data, where longer latencies are acceptable. These industrial monitoring applications, in general, do not need high data throughput or constant updating. Instead, emphasis is placed on low power consumption in order to maximize the lifetime of the battery-powered devices that make up the network.

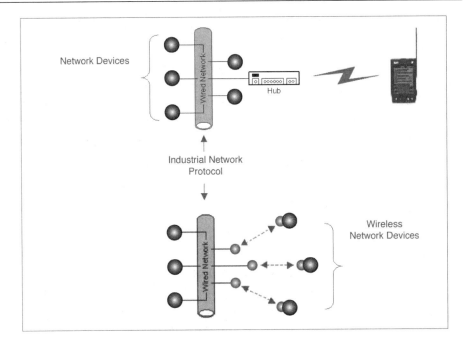

Figure 2–2:   Typical industrial use cases

Typical applications in industrial automation will include wireless access points operating as gateways of a wired industrial protocol (e.g., DeviceNet™, Fieldbus®, INCOM™, HART®, and others). These access points will allow monitoring and parameter configuration of devices connected to the wired network from a PDA (wireless hub) or will allow a wireless link to a network device (cable replacement). Figure 2–2 shows the industrial uses cases explained previously.

 In general, industrial applications do not have a particular network topology that characterizes them. The inherent network topology flexibility of IEEE Std 802.15.4 enables a wide range of applications in diverse industrial installations.

HOME AUTOMATION AND NETWORKING

The consumer and home automation market presents a significant potential because of its size. LR-WPAN devices will replace wires in home environment at a very low cost. Types of potential LR-WPAN devices include consumer electronic devices, personal computer peripherals, interactive toys and games, as well as home security, lighting control, and air conditioning systems [2]. Most of these applications have an industry group interested in using a low-cost, low data-rate wireless solution. Some of these industry groups are as follows: the

Consumer Electronics Association (CEA), HomePlug Alliance, and the ZigBee Alliance.

Of particular interest is the analysis performed by the ZigBee Alliance that identifies home automation applications as providing the lower bound on required data throughput, and the upper bound on acceptable message latency, for LR-WPANs.

 IEEE Std 802.15.4 adopted a maximum data rate of 250 kb/s as a result of an analysis performed for home automation applications.

Consumer Electronics

The consumer electronics sector includes radios, televisions, VCRs, CD and DVD players, remote controls, and appliances in general. In this sector, the first application may be the truly universal wireless remote control. The opportunities to start bundling products and services with a common control mechanism (the IEEE Std 802.15.4-enabled remote) will open up many interesting opportunities. With the convergence of PC and consumer electronics markets, IEEE Std 802.15.4 gives the end user the ability to control devices from both segments with the same intuitive device they are familiar with: the remote control. It will now be possible to have remote control units assigned to individuals instead of being assigned to a piece of equipment (in addition to reducing the many remote control units in the home). Individualization and two-way remote controls become possible, to enable feature rich programming and control. However, an often overlooked application is for communication links between devices to set volume levels, equalizer, and other control settings within a home entertainment center that may be distributed throughout the home.

PC Peripherals

PC peripherals, including wireless mice, keyboards, joysticks, PDAs, and games, represent a large segment inside the home for low-rate wireless connectivity. In addition, PC manufacturers and software companies are trying to change the interface to the PC, in order to transform it into a more consumer-oriented device. With the advent of remote controls and PDAs, this will become the way a PC is controlled. IEEE Std 802.15.4 devices will be able to provide the kind of low-power, low-cost solutions enabling this application scenario.

Home Automation

The home automation sector, which includes heating, ventilation, and air conditioning (HVAC), security, lighting, and the control of objects such as curtains, windows, doors, and locks, represents certainly a great opportunity for wireless innovation inside the home. In many cases, the thermostats in houses are located in the rooms where minimal time is spent (corridors, foyers, etc.). As a result, the temperature readings often do not correlate well with the temperature experienced by those living in the home, and hence, control of the environment in the home is often inefficient. Wireless thermostats could be dispersed throughout the home, and they could communicate with wireless vent controllers in the HVAC system to regulate the temperature in rooms individually. Window curtains could be drawn when the television is turned on during daylight hours; wall clocks could be coordinated with a master reference so that resetting after power failures would be automatic (no more blinking "12:00" on the VCR!). Home security could be as simple as the automotive remote keyless entry key fob—a single button press could lock all doors and windows; wireless smoke detectors and glass breakage sensors would connect to the home's security system. With the availability of all of these applications using the IEEE Std 802.15.4, a common remote control can be used to control all of these functions in addition to the consumer electronics devices.

Some manufacturers are planning to integrate IEEE Std 802.15.4 transceivers in light bulbs and ballasts. This would eliminate wiring (especially important for retrofit) and will enable standard wireless switches to be placed anywhere in the home without the need to break walls!

Home Security

Similarly to home automation, security sensors are dispersed around the home environment. These sensors are normally connected to a telephony-based backbone. Again, the value proposition of a wireless capability using LR-WPAN transceivers is based on the ease of installation.

There are two important challenges for wireless sensors in home security. The first one is related to power consumption of the sensor. Commonly, these sensors are not mains powered, because they are normally located on windows and doors. This challenge is overcome by IEEE Std 802.15.4, which as stated before, was designed with a simple yet efficient protocol stack. Furthermore, the medium access control (MAC) definition offers the upper layers an interface that allows full control to enable or disable the radio functionality. The second challenge is related to wireless range due to the propagation of radio frequency signals. The output power of IEEE Std 802.15.4 transceivers is bounded due to governmental regulations and by the need to conserve power (when necessary). IEEE Std 802.15.4 offers peer-to-peer protocol support that allows a network layer (not included in the standard) to implement a multihop network, effectively increasing the effective range.

Personal Healthcare

The personal health care sector includes sensors, monitors, and diagnostics. This is a field separate from medical telemetry; personal health care includes items such as pedometers and pulse rate monitors used by joggers and other athletes, and time and depth recorders used by scuba divers. These devices produce data that often need to be transferred to a display device; a wireless connection is often the only practical method to do this. Another application is health record maintenance; one's daily weight, temperature, and so on can be easily transmitted via a wireless link between the proper measuring device (scale, thermometer, etc.) and recorded on a personal computer or a PDA. With the availability of this information on the Internet, it is possible for health care professionals to access this data and monitor the status of persons remotely — no more need to drive to the weight loss center every week!

Toys and Games

Low-rate wireless networks have much to offer in the toys and games sector, especially when the communication is among toys or a toy and a personal computer. Toys are extremely cost-sensitive; the limiting factor on such computationally intensive tasks as voice recognition and synthesis in a toy often is not the technical feasibility of the required task, but the need to perform it for a low total hardware cost. By adding a wireless link between the toy and a nearby personal computer, the cost of the toy may be reduced, because the toy only needs to include the wireless link and the necessary sensors and actuators (e.g., microphones and speakers). Such "PC-enhanced" toys may exhibit sophisticated behaviors (e.g., mobile robots), limited only by the capabilities of the computer and the wireless link. Other opportunities in this sector include wireless gaming between individuals or groups; in addition to simple wireless links between players, both toys and games could receive "updates" when within range of a computer, and then operate autonomously in the new mode, or with the new personality or feature, when leaving the range of the computer. Wireless gaming manufacturers using IEEE Std 802.15.4-based solutions will be able to address the installed base of games in addition to the new games, as the cost and power consumption of a wireless add-on module is targeted to be low enough to make it economically enticing to users to add on to their existing games.

A summary of the performance requirements of these application sectors is shown in Table 2–1. Note that the required data rate for most applications within each sector is significantly below the maximum rate shown.

Table 2–1: Requirements of home networking market segments

Sector	Max Required Data Rate	Maximum Acceptable Message Latency
Consumer Electronics	3 kb/s	16.7 ms
PC Peripherals	115.2 kb/s	16.7 ms
Home Automation	10 kb/s	100 ms
Personal Healthcare	10 kb/s	30 ms
Toys and Games	115.2 kb/s	16.7 - 100 ms

AUTOMOTIVE SENSING

Wireless communication is finding its way into cars, as driver comfort and the number of features offered increase. The first wave of wireless technology in a car came from the "remote keyless entry" application and its derivatives. The second wave is happening now in the form of "cable replacement" for telematics applications. Bluetooth has taken the lead in addressing this type of application, maintaining a major focus on telephony applications such as handling of a mobile phone from embedded devices in a car (in-car hands-free speech) and car personalization through data exchanged from a mobile phone to the car.

In general, cost plays a fundamental role for vehicle applications not related to safety and luxury convenience. Today, a car is loaded with a great amount of sensors and actuators distributed all around the vehicle. The use of these devices has originated a significant growth in wiring, causing a great impact on the installation costs, diagnostics, maintenance, and even fuel consumption!

The wireless option introduces flexibility in installation and an advanced alternative to wired connections. A special challenge for automotive applications is meeting the harsh automotive environment while maintaining the low-cost requirements.

An application example in the category of "virtual wire" is a tire pressure monitoring system. The system consists of four pressure sensors, mounted on each tire, and a central station to receive the collected data. Because the pressure sensors have to be mounted onto the tires, this application does not permit the use of any communication wires or power cables. Therefore, sensors have to be battery powered. Because it is impractical to replace the sensors or their batteries between tire changes, it is required that the sensor batteries last at least three years, and preferably five years. This puts significant constraints on the power consumption of the electronic components and requires power management capabilities. The data that needs to be communicated (the measured tire pressure) are, in most cases, only a few bits in size. This information is transmitted about every 1 to 10 minutes under non-alarm conditions. Unless there is a fast loss of pressure, the message latency is not of significant concern. In case of a sudden pressure loss, the central control unit should be notified immediately; in which case, the power consumption is not of concern because most likely the tire has to be replaced. Extreme automotive environmental conditions and the metallic structure of the car complicate the RF design. In addition, the shape of the rim has a significant impact on the radiation pattern from the wireless sensor. To overcome this issue, repeater devices, which should not add a significant cost to the system, can be added to the network to increase communications reliability. Of course, an implicit advantage not present in actual applications is the possibility to have two-way communications with the sensor, thus increasing the reliability of the system.

Precision Agriculture

Another challenging application for LR-WPANs is precision agriculture, also known as *precision farming*. Precision agriculture is an environment-friendly system solution that optimizes product quality and quantity while minimizing cost, human intervention, and the variation caused by the unpredictability of nature. Today, agriculture is still both user and environmentally demanding. It is mainly hardware-oriented with manual and on-site control using independent dumb machines, which produce unpredictable quality and quantity. With the new paradigm of precision agriculture, farming would become more information and software-oriented, using automatic and remote-controlled, networked smart machines. This application requires large mesh-type networks consisting of potentially thousands of LR-WPAN devices linked with sensors. These sensors will gather field information such as soil moisture content, nitrogen concentration, and pH level. Weather sensors for measuring rainfall, temperature, humidity, and barometric pressure will also provide the farmer with valuable information. Each sensor will pass the measured data to its corresponding LR-WPAN device, which in turn will pass it through the network to a central collection device. In order for the sensor data to be useful, location-aware technology is necessary to correlate each sensor with its specific location in the field. The combined information will give farmers an early alert of potential problems and allow them to achieve higher crop yields.

The precision agriculture application is at the low end of the LR-WPAN application range, requiring the transmission of only a few bits of data per day by each deployed device. The data flow will be asynchronous in nature, with minimal restrictions on data latency. This combination of factors is advantageous for achieving long battery life. The challenge of this application is the topology of the network, because it requires a mesh topology where some devices serve as repeaters for others, relaying messages to the final destination, while still being power conscious to obtain the required usage life. The network should also be self-configuring because manual setup of a network of the proposed size is not feasible.

MISCELLANEOUS APPLICATIONS

A unique application scenario, falling within the consumer market, is a classroom calculator network. The teacher's workstation or the PAN coordinator would send tasks and math problems to each of the student's graphic calculators or the network devices. After completion, the students would upload their solutions back to the teacher's workstation. This network would need to support only a small number of devices, and it would require disallowing any peer-to-peer communication to prevent students from exchanging the solutions. The typical payload would be 100 to 500 bytes of information, sent several times per student per hour. It is desired that the batteries powering the calculator and communication function last the duration of a semester. Although this is definitely a higher throughput application than are the ones presented so far, it is still very well suited for an LR-WPAN.

Another important application in the category of "wireless hub" is remote metering and remote configuration. This type of applications enables improved hazard protection and operator convenience. Devices using an inexpensive LR-WPAN device will be able to exchange information wirelessly using PDAs, which can have previously programmed configuration information or contain software that will allow data to be synchronized with large databases.

Part II

Chapter 3 IEEE Std 802.15.4 Technical Overview

...warming up

The IEEE Std 802.15.4 Working Group established the goal of ultra-low complexity, low-cost, and extremely low-power wireless connectivity among inexpensive fixed, portable, and moving devices. With relaxed throughput and latency requirements, low-cost and low-power design could be achieved. The shared dream of "ubiquitous, untethered short-range communications" began to be realized.

Substantial costs can be associated with administration, sales, and marketing activities, as well as the cost of manufacturing and operation of wireless products. To achieve the goal of low total product cost as well as long battery life for low operating cost, the IEEE Std 802.15.4 provides reasonable tradeoffs in several performance metrics. To control the administrative costs of both implementer and user, IEEE Std 802.15.4 devices employ unlicensed radio bands. Because IEEE Std 802.15.4 service is designed to be a short-range service without infrastructure, and able to support very large networks, both transmit and receive functions can and must consume little power and have a low operating cost. This chapter presents some of the unique features of IEEE Std 802.15.4 that enable this performance vision to be met.

HOW LOW-COST AND LONG BATTERY LIFE ARE ACHIEVED

Duty Cycle

The battery is a relatively high-cost component in a transceiver system. The battery provides the energy for the communications, but at a cost of operation and replacement. Total system cost could not be minimized if a high-cost, exotic battery technology was required. The characteristics of low-cost batteries limit the choices for other attributes of the system. For example, the fully charged energy capacity and the instantaneous power delivery capacity of most batteries are interrelated. The fully charged energy capacity is achieved only if the instantaneous power delivery requirement is limited. Because of this limitation, to achieve long battery life, the energy must be taken continuously at an extremely low-rate, or in small amounts at a low-duty cycle. Due to the power consumption of practical wireless circuits, it is impractical to attain the desired battery life under constant operation; IEEE Std 802.15.4 was therefore designed to support very-low duty-cycle operation. The standard allows some devices to operate at low-duty cycles with both the transmitter and receiver inactive for over 99% of the time they are in operation.

However, in practical implementations there will always be a small amount of standby power consumed for timers, and so forth, when the wireless circuits are inactive. To reduce the time-average power consumption, both the active power and the standby power of an implementation should be reduced as much as possible. However, for a given technology and set of applications supported by the network, there is a practical limitation for both the active power and the standby power. For most applications, the active power is much larger than the standby power. Under this assumption, we can see that by reducing the duty cycle low-power consumption levels, and associated long battery lifetimes can be reached.

For example, for a device with 10 mW active power and 10 μW standby power, if the duty cycle is 0.1%, then the time-average power drain is about 19.99 μW. If the device is supplied by a 750 mAh AAA battery, linearly regulated to 1 V, it will have a battery life of more than 37,000 hours, or more than four years.

To support low-duty cycles, the IEEE Std 802.15.4 beacon packet can be as short as 544 μs in the 2.4 GHz band, while the superframe period (the time between network beacons) may be extended from 15.36 ms to over four minutes. This results in a beacon duty cycle that may be set from 2.3% to 0.000216% (2.16 ppm). Further, the standard supports a non-beacon mode in which a network may operate without any beacons. The non-beacon mode enables the slaves in a master-slave star network, for example, to remain in standby mode indefinitely, only contacting the master (which may be mains powered, perhaps, and therefore capable of constant reception) when an event occurs. The slaves may therefore have an almost unlimited battery life, limited primarily by their standby power consumption. The non-beacon mode, in fact, helps meet regulatory requirements for operation in the 868 MHz band, which has a maximum duty cycle limitation of 1%.

Modulation

With its low-cost and low-power consumption goals in mind, the IEEE Std 802.15.4 communications protocol was designed to support digital data communication only (i.e., analog service is not supported). Data-only service allows the modulation scheme to be chosen to be highly efficient, enabling a low-cost implementation. Likewise, the protocol supports only half-duplex operation so that the transmitter and receiver are not required to be active simultaneously.

The PHYs employ a spreading sequence to provide the benefits of Direct Sequence Spread Spectrum (DSSS) service; the 2.4 GHz PHY employs a form of multilevel orthogonal signaling, sending four bits per symbol, which simultaneously enables both a high-data rate (to return to standby mode quickly) and a relatively low-symbol rate (to minimize active power). In the 868/915 MHz PHY the chip modulation schemes are either raised-cosine-shaped Binary Phase Shift Keying (BPSK), half-sine-shaped Offset Quadrature Phase Shift Keying (O-QPSK), or root-raised-cosine-shaped Parallel Sequence Spread Spectrum (PSSS). The 2.4 GHz PHY uses a half-sine-shaped Offset Quadrature Phase Shift Keying (O-QPSK) chip modulation scheme. Further, the half-sine-shaped O-QPSK maintains a peak-to-average carrier power ratio of one, which minimizes both power consumption and implementation complexity.

Use of Direct Sequence Spread Spectrum

DSSS is one of several techniques to increase the bandwidth of a transmitted signal. The wideband techniques provide improved communications qualities but usually sacrifice spectrum utilization. For instance, the conventional broadcast FM service employs a form of spectrum spreading (wideband FM) that, although it does not employ DSSS, exemplifies the improved Signal-to-Noise Ratio (SNR) communications benefits possible with wideband techniques.

DSSS can be modeled as applying a prearranged pseudo-random digital sequence to directly phase-modulate the already data-modulated carrier, at a rate in excess of the data rate. The resulting DSSS signal occupies a much greater bandwidth, albeit with a lower spectral power density. The signal is recovered by demodulating the received signal with a replica of the same modulating pseudo-random digital sequence. The replica is generated in the receiver by a technique that ensures that the replica is closely coherent with the modulation in the transmitted signal. An ideally recreated replica transforms the received DSSS signal back into a replica of the transmitter's original data-modulated carrier. The DSSS process spreads the original signal into a wider bandwidth for transmission over the channel, and then despreads the signal at the receiver to recover the original signal and the information it contained.

The benefits of the DSSS form of spectrum spreading are a direct result of the relative coherence of the pseudo-random digital sequences used to phase-modulate and recover the replica of the original data-modulated carrier. Only those signals that are modulated by the correspondent coherent pseudo-random digital sequence are despread back into a replica of the original data-modulated carrier. All other signals are further spread by the receiver despreading phase-modulation process.

Because a spreading process reduces the power spectral density of noncoherent signals being modulated, interfering signals, adjacent channel signals, and even other noncoherent spread-spectrum devices on the same channel have a lowered spectral density at the receiver within the narrow bandwidth of the original data-modulated carrier. All other signals are spread to a wider bandwidth, lower spectral density, and smaller energy contribution when filtered to the original data-modulated carrier bandwidth. This "processing gain" can be used, for example, to reduce requirements on channel filters, thereby lowering implementation costs.

The cost benefits of DSSS result from an SNR improvement of coherent versus noncoherent signals, but there are other pertinent effects also.

One effect is that the power spectral density of the signals emitted from the spread spectrum transmitter are lowered relative to the power spectral density of the unspread data-modulated carrier, and are less likely to interfere with narrowband services using the same band. This coexistence effect enables the use of multiple services in the same bandwidth. A DSSS service can be added to existing spectrum allocation with reduced interference to existing services and likewise be relatively unaffected also by those services. DSSS is a nearly ideal means to share spectrum.

A more subtle advantage of DSSS is that its implementations may be made of largely digital circuits, with few analog circuits. Because digital circuits follow Moore's Law, and shrink with improving integrated circuit lithography, they, and the resulting transceiver implementation, become cheaper to produce over time.

Finally, because of these favorable characteristics of spread spectrum operation, use of DSSS (or another form of spread spectrum, frequency hopping) is mandated by many regulatory bodies worldwide for the unlicensed bands under their jurisdiction. IEEE Std 802.15.4 makes use of these spectrum allocations to define a practical low-power, low-cost, unlicensed service.

 Moore's Law establishes that the number of transistors on an integrated circuit doubles every two years.

Transmitter Power

Additionally, although IEEE Std 802.15.4 permits any legally acceptable output power, it requires only that the compliant device be capable of transmitting −3 dBm, well within the instantaneous power capacity of inexpensive battery sources, and within the capability of highly integrated and low-cost System-on-a-Chip (SoC) implementations.

Receiver Sensitivity

The required minimum −92 dBm RF sensitivity specification in the 868/915 MHz PHY, and the −85 dBm RF sensitivity specification in the 2.4 GHz PHY, permit the use of a simple receiver to achieve compliance. Simple, low-cost receiver designs, with little radio-frequency amplification (which can be expensive in terms of power consumption) are possible.

Quality of Service

To reduce implementation complexity, IEEE Std 802.15.4 does not support isochronous communication, and it does not support multiple classes of service within a single PAN. An exception is that the standard does support the optional use of Guaranteed Time Slots (GTS), which reserve network time for synchronous communication (avoiding potential channel access delays), at the discretion of the PAN coordinator. This extends the application space of IEEE Std 802.15.4 devices to include applications requiring low-latency communications, such as wireless joysticks and mice.

Although IEEE Std 802.15.4 is of unusually low complexity, the PHY layers support instantaneous link data rates of 20 kb/s, 40 kb/s, 100 kb/s, and 250 kb/s. The data rates are high relative to the data throughput in order to minimize device duty cycle.

In summary, IEEE Std 802.15.4 is designed to trade lower data throughput and higher message latencies for significantly lower cost and power consumption, while still providing useful communications for a wide class of services.

OTHER NOTABLE IEEE STD 802.15.4 FEATURES

Network Components

An IEEE Std 802.15.4 network is composed of a set of IEEE Std 802.15.4 wireless devices. Each network contains exactly one specialized central network coordinator, called the *PAN coordinator*. Only the PAN coordinator can establish a new network and it defines the structure and operating mode of the network. Other devices can join the network by requesting permission from the PAN coordinator. Besides the PAN coordinator, the standard distinguishes between two additional logical device types, one being a coordinator and the other, a network device. A coordinator is a device that can provide coordination services to other devices of the network, such as acting as a proxy for network devices that are not within range of the PAN coordinator. An 802.15.4 network consists of one PAN coordinator and at least one network device.

IEEE Std 802.15.4 defines two types of devices, the Full Function Device (FFD) and the Reduced Function Device (RFD). The FFD contains the complete set of MAC services and allows it to operate as any of the three possible personalities. The RFD contains a reduced set of the MAC services, and it can only operate as a network device.

 RFDs were defined to allow the implementation of extremely simple devices that require minimal resources in terms of processing power and memory capacity. These characteristics have a direct impact on the implementation of lower cost devices to enable applications such as light switches, stick-on sensors, and others.

Multiple Network Topologies

An IEEE Std 802.15.4 network may operate in one of two basic topologies. The first topology is called a *star* topology, and it is formed around a full-function device that is designated the PAN coordinator and that acts as a hub with a collection of additional full or reduced function devices that act as data terminal locations. In this regard, the PAN coordinator performs a special function in the network; it is the only device in the network that forms a direct link to more than one other device.

The second topology enables peer-to-peer communication without the direct involvement of a designated network coordinator as such (although a PAN coordinator is required somewhere in the network), and each device is able to form multiple direct links to other devices. The capability to form direct links with more than one other correspondent device ensures that the network graph can form redundant paths, but the complexity is increased. As a consequence of this second topology, many types of peer-to-peer network architectures can be supported.

One such complex network that can be supported by IEEE Std 802.15.4, but is not part of the specification, is the cluster-tree. A cluster-tree (see Figure 3–1) can be interpreted as a hierarchical tree of network device clusters. The cluster-tree arrangement provides complete connectivity, somewhat simplified routing, and a minimum number of direct links compared to a fully interconnected peer-to-peer structure, but it may result in increased data traffic latency.

The leaves of a cluster tree (those devices at the edge of the hierarchy) may be either FFDs or RFDs, since they do not need to relay messages for other devices. However, all other devices in the network must be FFDs. Exactly one device in the network assumes the special role of the PAN coordinator, which has a function much like the root device of a hierarchical tree. The PAN coordinator may be selected because it has special computation capability, a bridging capability to other network protocols, or simply because it was the first participant in the formation of the network. Other devices may function as roots of individual clusters; these are designated as cluster heads, or coordinators. The clusters form a hierarchical tree with typical parent-child relationship between devices. In this cluster-tree network example, all leaf nodes (devices at the end of the branches) are considered network devices, while all other devices need to be coordinators.

Networks topologies built on the IEEE Std 802.15.4 peer-to-peer communication feature are supported by the higher levels of the protocol stack which are outside the scope of IEEE Std 802.15.4. They are introduced here to show the flexibility of which the standard is capable; additional discussion is in Chapter 6.

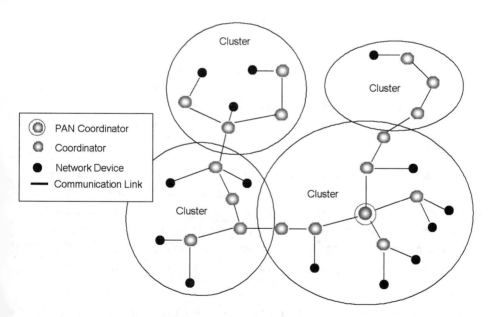

Figure 3–1:   Cluster-tree network

Channel Access

Regardless of the type of network employed, each network device employs a Carrier Sense Multiple Access-Collision Avoidance (CSMA-CA) protocol to avoid wasteful collisions when multiple simultaneous transmissions might otherwise occur. Exceptions to this are beacon transmissions, transmissions in GTSs, and acknowledgments, each of which is transmitted without the CSMA-CA protocol.

The CSMA-CA protocol is based on the shared nature of the RF channel. Whenever two or more senders are active on a channel simultaneously, the probability that any one of them is successful in sending a message decreases due to collisions and their mutual interference. In the RF environment, a great deal of the actual interference depends on the location of the competing transmitters and their respective correspondents, but the location information is often not available to the transmitting devices. One way to avoid collisions in the channel is to listen first and to transmit only if the channel is clear. The carrier sense improves the probability of access to the clear channel and the possibility of a collision is reduced. In this fashion, the channel capacity is more fully utilized.

Multiple PHYs

IEEE Std 802.15.4 defines a total of four PHYs—one mandatory and two optional PHYs for the 868 and 915 MHz bands, and one mandatory for the 2.4 GHz band. Note that the 868/915 MHz PHY requires that a compliant device be capable of operating in both the 868 MHz band and the 915 MHz band. This stipulation reflects the desire to minimize the number of potentially incompatible products on the market, and it recognizes that there is little cost penalty in this requirement, due to the small frequency difference between the two bands.

The Task Group that developed IEEE Std 802.15.4 was comprised of volunteers from around the world, and therefore considered that compliant devices may be developed and used worldwide. To keep administrative costs low, IEEE Std 802.15.4 assumes that all devices will perform in unlicensed bands but each would fall under the provisions of the regulatory agency with jurisdiction for the particular service area. Unfortunately, the regulatory agencies worldwide have allocated the spectrum differently between and among themselves. There are three bands available for this type of service—one worldwide and two regional special cases.

For implementers, the choice of bands depends on more than simple technical considerations. The 868 and 915 MHz bands, although they are likely to be less crowded and may offer better QoS, are not available worldwide. The decision to implement products for a limited market, albeit with possible better service, falls in the realm of marketing and business rather than purely technical matters. There are "hidden" costs of distribution and marketing of otherwise identical products that differ only in frequency band. There are interference problems that bring up questions of how to keep products from moving from one regulatory region to another. Other applications (like wireless luggage tags) do not fit into a regional marketing strategy and do not present a consideration, but the possibility of regional strategies exists for some products.

Portions of the 2.4 GHz band are available nearly worldwide; the 868 MHz band is available in Europe, and portions of the 915 MHz band is available in North America, Australia, New Zealand, and some parts of South America. IEEE Std 802.15.4 defines a single channel with a mandatory raw data rate of 20 kb/s and two optional data rates of 100 kb/s and 250 kb/s in the 868 MHz band. In the 902 — 928 MHz band, the standard defines 10 channels with a mandatory raw data rate of 40 kb/s (doubled chip rate with respect to the 868 MHz band) and an optional raw data rate of 250 kb/s. Higher in the spectrum, at 2.4 GHz, IEEE Std 802.15.4 defines 16 channels with a raw data rate of 250 kb/s.

Error Control

IEEE Std 802.15.4 employs a simple full-handshake protocol to ensure reliable data transfer and good QoS. With the exception of broadcast frames (e.g., beacons) and the acknowledgment frame, each received frame may be acknowledged to assure the transmitting device that its message was, in fact, received. If a requested acknowledgment frame is not received by the transmitting device, the entire transmitted frame may be repeated.

To detect that a message has been received correctly, a cyclic redundancy check (CRC) is used. The message bits are treated as a long binary number and divided by a relatively large prime number. The quotient of the division is discarded, and the remainder of the division is transmitted along with the message. The same division with the same prime number is performed at the receiver, and a match signifies a high probability of an uncorrupted communication.

FOUR FRAME TYPES

There are four frame structures, each designated as a PHY Service Data Unit (PSDU) in the standard for data transactions—a beacon frame, a data frame, an acknowledgment frame, and a MAC command frame. As shown in Figure 3–2, all frames are structured in a very similar fashion, with the primary differences in their purpose or payloads. Each PHY Protocol Data Unit (PPDU) is constructed with a Synchronization Header (SHR), a PHY Header (PHR), and a PHY Service

Data Unit, composed of the MAC Protocol Data Unit (MPDU) as a data structure that services the MAC protocol layer. The MPDU is constructed with a MAC Header (MHR), a MAC Footer (MFR), and a MAC Service Data Unit (MSDU)—excepting the MPDU of the acknowledgment frame, which does not contain an MSDU.

The MSDU is a data (i.e., payload) field component of a given frame containing information pertinent to the MAC services supported by the frame, including superframe identification and sequencing information, addressing information, and other information.

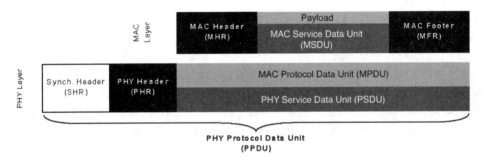

Figure 3–2:   Frame structures format

Beacon Frames

Transmission of the beacon frame is available only to FFDs in the network, regardless of whether the network is a star, cluster-tree, or other topology. The beacon frame is provided as a service originating in the MAC protocol layer and interfaced to the PHY protocol layer. It has a number of uses, including superframe boundary marker, frame synchronizing signal, and association supervision, all as a service to the higher protocol layers.

As a superframe boundary marker, the beacon frame provides a timing reference to designate the boundaries and structure of a superframe. In the higher level of the protocol, the superframe permits a fixed number of frames to be designated between beacon frames, whether each frame position is occupied or not.

As a frame synchronization signal, the beacon frame allows the synchronization of the superframe to a known starting time, preventing collisions, allowing receivers to sleep whenever there are no transactions, and permitting good message latency by limiting the typical delay of traffic to the length of one super-frame on any one link. Further, the synchronization allows the network device transmitting the beacon to become the timing reference for link communication and allows a relaxation of an individual network device's timing requirements.

Each network device may have a less precise timing reference that is periodically adjusted to match the beacon device's reference.

Data Frames

The data frame is available to all devices in any network, regardless of whether the network is a star, cluster-tree, or other topology. The data frame, like the beacon frame, is provided as a service in the MAC layer due to a request from a higher layer. It is interfaced to the PHY layer. It provides the primary data payload as a service to the higher protocol layers.

Acknowledgment Frames

The acknowledgment frame is also available to all devices in any network, regardless of whether the network is a star, cluster-tree, or other topology. The acknowledgment likewise is provided as a service originating in the MAC protocol layer and interfaced to the PHY protocol layer. It provides only the acknowledgment for receipt of data as a service to the higher-level protocol layers for end-to-end message control. The acknowledgment frame does not contain a MAC payload.

MAC Command Frame

The MAC command frame is also available to all devices in any network, regardless of whether the network is a star, cluster-tree, or other topology. The MAC command frame, like the other frames, is provided as a service originating in the MAC protocol layer and interfaced to the PHY protocol layer. It provides the primary supervisory payload as a service to the MAC protocol layer.

For each of these four frame types, the MPDU is further encapsulated with an additional preamble and PHY header to form the physical layer protocol data unit (PPDU). This complete data structure allows a receiver to achieve symbol synchronization with the frame and obtain the frame length information.

Chapter 4 Physical Layer

...from bytes to watts

The physical layer (PHY) provides the interface with the physical medium where the actual communications occurs. The PHY layer is the lowest component in the ISO/OSI model and is in charge of providing control (activation and deactivation) of the radio transceiver, energy detection, link quality, clear channel assessment, channel selection, and the transmission and reception of message packets through the physical medium [26].

REGULATORY CONSIDERATIONS

Governments around the world regulate and administer the radio-frequency spectrum. In most cases, there are band allocations for unlicensed operation, given that the manufacturer can ensure operation within some pre-established limits in output power, duty cycle, modulation, and other parameters. Station licenses, along with an identification call sign, are issued to transmitters for many of the allowed services, but there is also a provision for the operation of some devices that do not require station licenses. The particulars of unlicensed services vary throughout the world, but generally, the devices must conform to a set of regulatory limitations that are specific for each operation. Transmitters for each operation must pass a testing protocol to ensure that they meet the specific requirements. Once a particular design has passed the regimen of testing, it is issued a certification or type approval for that service. IEEE Std 802.15.4 devices are targeted toward type approval.

Implementations of IEEE Std 802.15.4 must conform to local regulations of the country where it will operate. In the case of Europe, Japan, Canada, and the United States, the regulation consists of unlicensed but type-approved DSSS services. It is the constraint for unlicensed but type-approved compliance that dictates the frequency bands of operation, as well as a few of the other characteristics of the service.

IEEE Std 802.15.4 is written so that conforming devices can be manufactured to operate in any of three particular bands. Two of these bands are limited to specific geographical regions, but one band is available for nearly worldwide service. Details of these bands are presented next.

European Regional Case

In Europe, the European Telecommunications Standards Institute (ETSI) has published recommendations that are recognized generally by all regulatory agencies, but each service area falls under individual national type-approval authority. Within the European service area, there is one common band with operation allowed between 868.0 and 868.6 MHz that supports a single channel of low data rate service with less than 1% transmission duty cycle [4], [5]. Again, this band is unique to European service, and regulations elsewhere preclude operation of unlicensed devices in this band.

IEEE Std 802.15.4 868/915 MHz PHY requires that a compliant device be capable of operation on both the 868 and 915 MHz bands. For the purposes of the standard, they are considered to be a single, contiguous band.

 In May 2002, the Electronic Communications Committee (ECC) within the European Conference of Postal and Telecommunications Administrations (CEPT) released their strategic plan for the future use of the frequency band 862–870 MHz. This plan delineates a roadmap to increase the frequency band available for spread spectrum devices. The allocated band would operate in the 865–868 MHz band, allowing the addition of at least three radio channels to a future release of IEEE Std 802.15.4.

North American Regional Case

Within the United States, the national regulatory agency is the Federal Communications Commission (FCC). The FCC has specific authority only within the United States; however, the FCC regulations are used as a model by many other nations in the Americas and the Pacific Rim. Just as in Europe, there is a unique band available for service; this band is called the 915 MHz band, and it covers the range between 902 and 928 MHz. This band enables the provision of ten channels of low data rate service. With some exceptions, unlicensed operation in this band is unique to North America, and regulations elsewhere preclude operation of unlicensed devices in this band.

Worldwide Case

To obtain economies of scale in product design, marketing, and distribution, and to enable applications that may require roaming between different regulatory regions, it is desirable to employ a single band that is available on a nearly worldwide basis. The ideal band would be unlicensed, have sufficient width to enable the use of many channels, and be high enough in the spectrum so that relatively efficient antennas are possible without being so high that single-chip implementations employing low-cost integrated circuit processes are precluded.

The band selected as best meeting these requirements is the 2.4 GHz Industrial, Scientific, and Medical (ISM) band, which extends from 2400 to 2483.5 MHz. This band, with very few exceptions, is available worldwide without licensing [6], [22], [23] and, with a wavelength of 12.25 cm, enables reasonably efficient yet physically small antennas. Further, it is compatible with modern silicon integrated circuit processes, and the wider bandwidth available enables the provision of 16 channels of high data rate service, allowing independent networks of similar devices to coexist without interfering with each other.

IEEE Std 802.15.4 was designed for regulatory compliance in each of these bands. For example, in the 868 MHz band, the product duty cycle is limited, therefore to meet this requirement, the non-beacon mode of IEEE Std 802.15.4 may be used to minimize the duty cycle of devices in this band. In many cases, some type of spread spectrum operation is required for operation in unlicensed bands. The use of DSSS in IEEE Std 802.15.4 allows this regulatory requirement to be met, while also enabling a low-cost product implementation with good transmission range.

FREQUENCY BANDS AND DATA RATES

In response to the regulatory availability of the three bands for unlicensed operation, IEEE Std 802.15.4 specifies technical objectives for operation within each band:

- *868–868.6 MHz Band:* This unlicensed band is available in most European countries for a 20 kb/s BPSK and optional 100 kb/s O-QPSK service as well as an optional 250 kb/s Parallel Sequence Spread Spectrum (PSSS) service. IEEE Std 802.15.4 also refers to this band as the 868 MHz band.

- *902–928 MHz Band:* Some portions of this unlicensed band are available in North America, Australia, New Zealand, and some countries in South America for 40 kb/s BPSK and optional 250 kb/s O-QPSK service as well as an optional 250 kb/s PSSS service. IEEE Std 802.15.4 also refers to this band as the "915 MHz" band (which represents the middle frequency of the band).

> The IEEE Std 802.15.4 868/915 MHz PHY requires that a compliant device be capable of operation on both the 868 and 915 MHz bands providing a 20 kb/s and 40 kb/s data service respectively. For the purposes of the standard, they are considered to be a single, contiguous band.

- *2.4000–2.4835 GHz Band:* This third unlicensed band is available in most countries worldwide for the 250 kb/s O-QPSK service. This band is referred to as the 2.4 GHz band.

Due to its nearly world-wide availability, the 2.4 GHz PHY may be the first choice for many IEEE Std 802.15.4 applications, especially those involving products, that entail travel between regulatory regions. Even for non-mobile applications, the 2.4 GHz band offers advantages of scale, distribution, and marketing: a single product can be sold in multiple locations around the globe, without concern for the regulatory region in which it will be used. This drives down production costs, while eliminating the supply chain expense of tracking multiple products to multiple destinations.

IEEE 802.15.4 2006 A revision of the original IEEE 802.15.4 standard, completed in 2006, introduced 2 additional modulating options in the lower frequency band. The new PHY layer options provide a higher data rate compared to what was previously provided.

However, for these same reasons, the 2.4 GHz band is used by many other services, from microwave ovens to WPANs and WLANs. This can result in congestion that is unacceptable for some applications and markets. To address this issue, the IEEE Std 802.15.4 868/915 MHz bands are available as an alternative. This feature allows system designers to use these regional bands in order to avoid the potentially crowded 2.4 GHz band. The regional bands may be a good design choice for applications such as utility meter reading that are inherently regional in nature and have limited device mobility. Similarly, diverse sensors in different industrial segments can make use of this feature. Additionally, the new optional higher data rates make these low-band PHYs more attractive.

Data Rates

Due to the physical characteristics of each band and the regulations where they are used, IEEE Std 802.15.4 specifies different data rates and modulations for the three bands used in the two PHYs. Table 4–1 shows the data rates (bit and symbol) and the modulation parameters specified in each band.

IEEE 802.15.4 2006 Any device operating in the 868/915 MHz bands must at least support a data rate of 20 kb/s or 40 kb/s using the specified modulation, respectively. Support for higher data rates using either modulation is optional.

The mandatory base data rate in the 868 MHz band is 20 kb/s but implementers may choose adding either an optional higher data rate of 100 kb/s using O-QPSK modulation or 250 kb/s using PSSS modulation. Similarly, the mandatory base data rate in the 915 MHz is 40 kb/s with implementers able to provide an optional rate of 250 kb/s using either O-QPSK or PSSS modulation.

Table 4–1: IEEE Std 802.15.4 frequency band and modulation parameters

Band	Frequency Band	Bit Rate	Symbol Rate	DSSS Spreading Parameters	
				Modulation	Chip Rate
868 MHz	868 - 868.6 MHz	20 kb/s	20 ksymbols/s	Binary Phase Shift Keying (BPSK)	300 kchip/s
		100 kb/s	25 ksymbols/s	Offset Quadrature Phase Shift Keying (O-QPSK)	400 kchip/s
		250 kb/s	12.5 ksymbols/s	Parallel Sequence Spread Spectrum (PSSS)	400 kchip/s
915 MHz	902 - 928 MHz	40 kb/s	40 ksymbols/s	Binary Phase Shift Keying (BPSK)	600 kchip/s
		250 kb/s	62.5 ksymbols/s	Offset Quadrature Phase Shift Keying (O-QPSK)	1 Mchip/s
		250 kb/s	50 ksymbol/s	Parallel Sequence Spread Spectrum (PSSS)	1.6 Mchip/s
2.4 GHz	2.4 - 2.4835 GHz	250 kb/s	62.5 ksymbols/s	Offset Quadrature Phase Shift Keying (O-QPSK)	2 Mchip/s

A more detailed explanation of the modulation parameters will be introduced in the following band specification sections.

CHANNEL ASSIGNMENT

The IEEE Std 802.15.4-2006 uses a combination of channel numbers and channel pages to specify the operating frequency of a radio. There are a total of 32 possible pages; only three of them are defined in the current standard, with the rest reserved for future expansion. Each page is logically divided into 27 channels, numbered 0 to 26. The pages are numbered 0 to 31. Page 0 contains one channel in the 868 MHz band, ten channels in the 915 MHz band, and sixteen channels in the 2.4 GHz band. The center frequencies of the channels for page 0 are shown in Table 4–2. Channel page 1 contains the channel assignment for the optional PSSS high data-rate service in the lower frequency bands while channel page 2 specifies the channel assignment for the optional O-QPSK physical layer service in the same bands. Both pages 1 and 2 shown in Table 4–3 contain one channel for operation in the 868MHz band and 10 channels for operation in the 915 MHz band, with the remaining 15 channels being reserved.

> **IEEE 802.15.4 2006** The concept of channel pages was introduced in the 2006 revision of the standard to accommodate the new physical layer options and to allow for future expansion. The original 27 channels specified in IEEE Std 802.15.4-2003 are now accessed through page 0, while each of the two new modulation mechanism operate on their respective individual channel page.

Table 4–2: IEEE Std 802.15.4 channel assignment for page 0

	Channel	Center Frequency (MHz)	Availability
868 MHz Band	0	868.3	
915 MHz Band	1	906	
	2	908	
	3	910	
	4	912	
	5	914	
	6	916	
	7	918	
	8	920	
	9	922	
	10	924	
2.4 GHz Band	11	2405	
	12	2410	
	13	2415	
	14	2420	
	15	2425	
	16	2430	
	17	2435	
	18	2440	
	19	2445	
	20	2450	
	21	2455	
	22	2460	
	23	2465	
	24	2470	
	25	2475	
	26	2480	

NEW OPTIONAL PHYSICAL LAYERS

While the lower data rates in the 868/915 MHz band have the advantage of improved range performance, the result is lower data throughput with respect to the 2.4 GHz–band data rate. This can be a particular challenge for larger networks operating in these bands. For the European 868

IEEE 802.15.4 2006 The 2006 revision of the IEEE 802.15.4 standard introduced two new physical layer options for the lower frequency bands that increase the throughput to equivalent levels of the 2.4 GHz band.

MHz band this challenge is emphasized since the data throughput is limited not only by the 20 kb/s data rate (itself a result of the limited available bandwidth), but also by the regulatory duty cycle limitation. A device operating in the 868 MHz

Table 4–3: IEEE Std 802.15.4 channel assignment for page 1 and 2

	Channel	Center Frequency (MHz)	Availability
868 MHz Band	0	868.3	
915 MHz Band	1	906	
	2	908	
	3	910	
	4	912	
	5	914	
	6	916	
	7	918	
	8	920	
	9	922	
	10	924	

band is allowed to transmit with only a 1% duty cycle. To alleviate this disadvantage, the 2006 revision of the original standard added two new physical layer options providing higher data rates in the lower frequency bands. The two optional PHYs provide more choices to the implementer and user. They also allow original equipment manufacturers certain performance tradeoffs based on application requirements when choosing a solution.

The two new optional physical layers include an O-QPSK PHY with a data rate of 100 kb/s at 868 MHz and 250 kb/s at 915 MHz, and also a PSSS PHY with a data rate of 250 kb/s in both of these bands. However, for backward compatibility with existing devices and for interoperability among new devices, the standard requires implementers of either option to also support the original 868/915 MHz physical layer that was introduced by the original 2003 version of the standard.

The optional O-QPSK PHY specification for the 868/915 MHz band is a derivative of the modulation used also in the 2.4 GHz band, allowing implementers to share design similarities with existing radios and enable a potential three-band transceiver. The optional PSSS PHY offers the same data rate at both low-band frequencies and uses a more complex modulation scheme but has the advantage of providing improved multipath performance.

PHY BIT LEVEL COMMUNICATION

The physical layer of the IEEE Std 802.15.4 protocol is responsible for the establishment of the RF link between two devices. It also provides for bit modulation, demodulation, and synchronization between the transmitter and receiver. Finally, the PHY also handles the packet level synchronization.

IEEE Std 802.15.4 specifies four different data rates that can be used in the three different bands. A device compliant with the 868/915 MHz PHY layer specifications of IEEE Std 802.15.4-2006 must also support the two lowest rates of 20kb/s and 40kb/s using the BPSK modulation scheme. Optionally, the device may also provide a 100 kb/s or 250 kb/s data service using either O-QPSK modulation or a 250 kb/s data service using PSSS modulation. The 2.4 GHz PHY layer provides a data rate of 250 kb/s using an O-QPSK modulation scheme with an m-ary quasi-orthogonal modulation technique.

2.4 GHz Band Specification

The 16-ary quasi-orthogonal modulation technique specified in the IEEE Std 802.15.4 2.4 GHz PHY layer is a method of data modulation that utilizes a particular 32-chip, pseudo-random sequence to represent four bits and simultaneously accomplish the spreading modulation. The data modulation is performed by means of cyclic rotation and/or conjugation (inversion of chips with odd indices) of the sequence. The pseudo-random sequence is started in different places, depending on the modulating data, transmitting four bits in each symbol period.

While five bits could be transmitted by the choice of 32 chips, four were chosen for the 2.4 GHz PHY to minimize implementation complexity. The transmitted 32-chip pseudo-random sequence is allowed to start only at every fourth chip of the sequence. Symbols 0–7 represent cyclic shifts in multiples of four chips. Symbols 8–15 use the same shifts as symbols 0–7, respectively, but use the conjugated sequence (i.e., the odd-indexed chips are inverted).

The IEEE Std 802.15.4 2.4 GHz PHY layer specifies a symbol rate of 62.5 ksymbols per second with four bits in each symbol; therefore, 250 kb/s service is attained. The 32-chip pseudo-random sequence to be transmitted is split between the orthogonal I and Q channels of the O-QPSK modulator, with the even-indexed chips placed on the I channel and the odd-indexed chips placed on the Q channel. A one-half chip delay is placed in the Q channel, creating the offset for O-QPSK. Because 32 (now complex) chips are transmitted in one symbol time (16 μs), the overall chip rate is 2 Mc/s. The chip rate in either I or Q channel, however, is 1 Mc/s.

Initially two distinct pseudo-noise (PN) sequences were considered for the I and Q channels of the quasi-orthogonal modulation in the O-QPSK, but the two were later combined to use a single PN sequence separated into even and odd bits between the I and Q channels. The advantage of a single PN sequence lies in that a single correlator is required that can be shared between the I and Q channels, thereby reducing implementation complexity.

The bit level processing consists of assembling four bits into a symbol, converting that symbol to a cyclically rotated 32-chip sequence as shown in Table 4–4, and modulating that chip sequence on the I or Q channel, respectively. The process is diagrammed in Figure 4–1.

The resulting combination of complex modulation, including the offset between I and Q channels is shown in Figure 4–2.

Table 4–4:   Symbol to chip mapping

Data Symbol (decimal)	Data Symbol (binary) (b3 b2 b1 b0)	Chip Values (c0 c1... C30 C31)
0	0000	11011001110000110101001000101110
1	0001	11101101100111000011010100100010
2	0010	00101110110110011100001101010010
3	0011	00100010111011011001110000110101
4	0100	01010010001011101101100111000011
5	0101	00110101001000101110110110011100
6	0110	11000011010100100010111011011001
7	0111	10011100001101010010001011101101
8	1000	10001100100101110000001110111011
9	1001	10111000110010010110000001110111
10	1010	01111011100011001001011000000111
11	1011	01110111101110001100100101100000
12	1100	00000111011110111000110010010110
13	1101	01100000011101111011100011001001
14	1110	10010110000011101111011100011000
15	1111	11001001011000000111011110111000

Because 32 chips are transmitted in one symbol time (16 µs), the overall chip rate is 2 Mc/s, and the length of a chip is Tc = 0.5 µs. Successive chips in each of the I and Q channels start every 2 Tc.

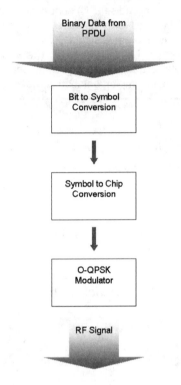

Figure 4–1: 2.4 GHz modulation and spreading

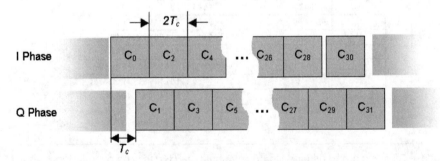

Figure 4–2: O-QPSK chip offsets

868/915 MHz Band Specification

The 868/915 MHz PHY layer based on the 2006 revision allows for multiple data rates using different modulation schemes thus providing OEMs with the opportunity to trade off certain features and performance characteristics depending on application requirements. To allow for backward compatibility, all devices compliant with the 868/915 MHz PHY layer specifications of IEEE Std 802.15.4-2006 are required to support the 20 kb/s and 40 kb/s data services of the 2003 version of the standard using the respective spreading parameters. This "base mode" of operation allows new devices to be able to communicate with IEEE Std 802.15.4-2003 compliant devices that are already deployed in the field and allows new devices that employ the new PHY layer options to interoperate. New applications or deployments without legacy devices may take advantage of the higher data rates and performance characteristics that the new physical layer options provide.

Required BPSK Mode Specification

The "base mode" of the 868/915 MHz PHY layer uses DSSS employing BPSK, providing a data rate of 20 kb/s in the 868 MHz band and 40 kb/s in the 915 MHz band. A complex signal path is not required for this mode.

The 868/915 MHz PHY specifies differential encoding of the transmitted data bits. If the raw data bit is "0," the BPSK data bit is transmitted in the same phase as the previous BPSK data bit, while if the raw data bit is "1," the BPSK data bit is transmitted in phase opposite to the previous BPSK bit.

The 868/915 MHz PHY specifies "conventional" DSSS, in which a single, 15-chip, pseudo-random sequence is transmitted in a symbol period to represent a "1," and the inverse of the sequence is transmitted to represent a "0." This process is illustrated in Figure 4–3. The chip rate is specified as 300 kc/s in the 868 MHz band, for a data rate of 20 kb/s; in the 915 MHz band, the specified chip rate of 600 kc/s enables a data rate of 40 kb/s. The reception process used to recover the transmitted data is shown in Figure 4–4.

The 868/915 MHz modulation and spreading process is illustrated in Figure 4–5.

 The BPSK modulation utilizes a different PN sequence, but it shares that sequence between the two lower data rates. The difference in data rates is due to the difference in chip rates, but the number of chip bits per data bit and the bit pattern are the same between the two.

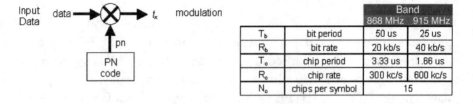

		Band	
		868 MHz	915 MHz
T_b	bit period	50 us	25 us
R_b	bit rate	20 kb/s	40 kb/s
T_c	chip period	3.33 us	1.66 us
R_c	chip rate	300 kc/s	600 kc/s
N_c	chips per symbol	15	

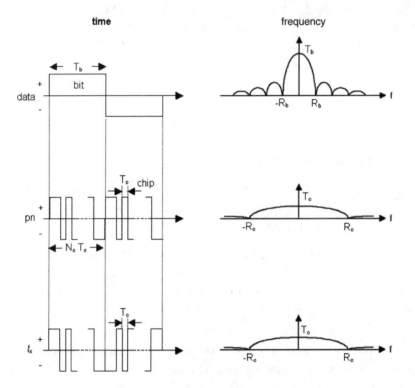

Figure 4–3:   DSSS modulation

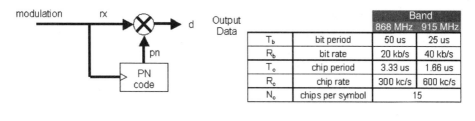

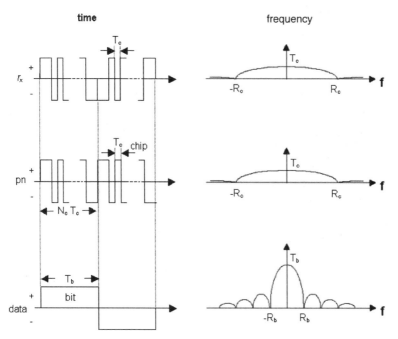

Figure 4–4: DSSS demodulation

Optional O-QPSK Mode Specification

The optional O-QPSK modulation scheme for the lower frequency bands is a derivative of the modulation used in the 2.4 GHz physical layer, and provides 100 kb/s data service at 868 MHz, and 250 kb/s data service at the 915 MHz band. It allows a significant improvement in data throughput over the base 868/915 MHz band BPSK mode while

The new optional O-QPSK PHY offers increased data throughput compared to the base mode while allowing implementers to share design similarities with existing transceivers operating in the 2.4 GHz band.

IEEE 802.15.4

2006

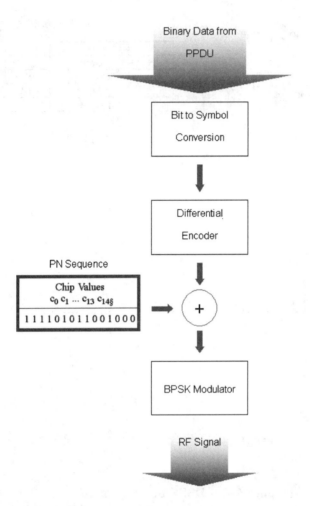

Figure 4–5: 868/915 MHz modulation and spreading

allowing implementers to leverage design similarities with the high band PHY. It employs a 16-ary quasi-orthogonal modulation scheme with a nearly orthogonal, 16-chip, pseudo-random noise sequence to modulate four data bits into each symbol. Identical to the 2.4 GHz PHY, the data modulation is performed by a cyclic rotation and/or conjugation of the chip sequence. Symbols 0–7 represent cyclic shifts in multiples of two chips. Symbols 8–15 use the same shifts as symbols 0–7, respectively, but use the conjugated sequence (i.e., the odd-indexed chips are inverted). The chip sequences used for this modulation scheme are shown in Table 4–5.

Table 4–5: Symbol to chip mapping for optional O-QPSK modulation

Data Symbol (decimal)	Data Symbol (binary) (b3 b2 b1 b0)	Chip Values (c0 c1 ... c14 c15)
0	0000	0011111000100101
1	0001	0100111110001001
2	0010	0101001111100010
3	0011	1001010011111000
4	0100	0010010100111110
5	0101	1000100101001111
6	0110	1110001001010011
7	0111	1111100010010100
8	1000	0110101101110000
9	1001	0001101011011100
10	1010	0000110101101 11
11	1011	1100000110101101
12	1100	0111000001101011
13	1101	1101110000011010
14	1110	1011011100000110
15	1111	1010110111000001

The symbol rate for the optional O-QPSK PHY is 25 ksymbol/s with a chip rate of 400 kchip/s at the 868 MHz band and 62.5 ksymbol/s with a chip rate of 1 Mchip/s at the 915 MHz band resulting in the previously mentioned 100 kb/s and 250 kb/s data services respectively. Before transmission, the symbol is split between the I and Q channels of the modulator. The even-indexed chips are put into the I-channel while the odd-indexed chips are put in the Q-channel. The transmission of the I- and Q-channel are offset by half a chip period.

The modulation process is identical to the one of the 2.4 GHz physical layer shown in Figure 4–1. The resulting complex modulation and its offset is shown in Figure 4–6.

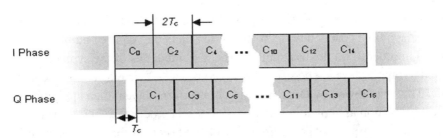

Figure 4–6: O-QPSK chip offsets for optional 868/915 MHz PHY

Optional PSSS Mode Specification

> **IEEE 802.15.4**
> **2006**
> The new optional PSSS PHY offers increased data throughput compared to the base mode and improved performance characteristics as result of the advanced multicoded modulation scheme.

The optional Parallel Sequence Spread Spectrum (PSSS) physical layer uses a multicode modulation scheme [28] based on amplitude shift keying (ASK). The advantage of this PSSS mode PHY is that it provides an improved multipath performance compared to the other two low-band physical layers while also achieving the same data rate available at the 2.4 GHz PHY in the 868 and 915 MHz bands. However, the drawback is a slightly more complex transceiver design compared to the other options.

The PSSS mode PHY uses a 31-chip base sequence. For the 868 MHz band this sequence is rotated by 1.5 chips and a one chip cyclic extension is added, which results in 20 nearly orthogonal pseudo-random sequences that are 64 half-chips long. The PSSS code table used for the 868 MHz band is show in Table 4–6. For the 915 MHz band the sequence is cyclically shifted by six chips with a one chip extension added. This operation results in five nearly orthogonal pseudo-random sequences that are 32 chips long. The PSSS code sequence used in the 915 MHz band is shown in Table 4–7. The modulation process consists taking the binary

Table 4–7:   PSSS Code Table for 915 MHz band

Sequence Number	Chip Number																															
	0	1	2	3	4	5	6	7	8	9	10	11	12	13	14	15	16	17	18	19	20	21	22	23	24	25	26	27	28	29	30	31
0	-1	-1	-1	-1	1	-1	-1	1	-1	1	1	1	-1	-1	1	1	1	1	1	-1	-1	-1	1	1	-1	1	1	-1	1	-1	1	-1
1	1	1	-1	1	-1	1	-1	-1	-1	-1	1	-1	-1	1	-1	1	1	-1	-1	1	1	1	1	1	-1	-1	-1	1	1	-1	1	1
2	-1	-1	1	1	-1	1	1	1	-1	1	1	-1	1	-1	-1	-1	-1	1	-1	-1	1	-1	1	1	-1	-1	1	1	1	1	-1	-1
3	1	1	1	1	1	-1	-1	-1	1	1	-1	1	1	1	1	-1	1	-1	1	-1	-1	-1	1	-1	-1	1	1	1	1	-1	-1	1
4	1	-1	1	1	-1	-1	1	1	1	1	1	1	-1	-1	-1	1	1	-1	1	1	1	1	-1	1	-1	-1	-1	-1	1	-1	-1	1

data of the PHY header and PHY payload and converting 20 of these data bits for the 868 MHz band and five of these data bits for the 915 MHz band into symbols. The last symbol is padded with "0" bits if less than the required number of bits remain in order to fill the last symbol. The individual bits of the symbol are converted to bipolar levels where a "1" bit becomes a "+1" and a "0" bit becomes a "–1." This is followed by the symbol to chip conversion and then applying the ASK modulation as shown in Figure 4–7. For this process the bipolar version of the first bit of the symbol is multiplied by the first sequence in the respective PSSS code table, the bipolar version of the second bit of the symbol is multiplied with the second sequence in the PSSS code table, and so on. The effect is that, depending on the data bit, the respective PSSS code sequence is either inverted or not.

Table 4–6: PSSS Code Table for 868MHz band

Chip Number

Sequence Number	0	1	2	3	4	5	6	7	8	9	10	11	12	13	14	15	16	17	18	19	20	21	22	23	24	25	26	27	28	29	30	31
0																																
1																																
2																																
3																																
4																																
5																																
6																																
7																																
8																																
9																																
10																																
11																																
12																																
13																																
14																																
15																																
16																																
17																																
18																																
19																																

Half-chip Number: 0 1 2 3 4 5 6 7 8 9 10 11 12 13 14 15 16 17 18 19 20 21 22 23 24 25 26 27 28 29 30 31 32 33 34 35 36 37 38 39 40 41 42 43 44 45 46 47 48 49 50 51 52 53 54 55 56 57 58 59 60 61 62 63

The products of these multiplications (20 multiplication results when operating in the 868 MHz band and five multiplication results when operating in the 915 MHz band) are summed to a single multilevel modulation sequence. This is followed by a two-step precoding process to eliminate the DC content of the multilevel modulated sequence and to normalize the sequence. To transmit the sequence over the medium, it is modulated onto a carrier using amplitude shift keying (ASK).

The synchronization header is transmitted using BPSK modulation at the same chip rate and pulse shaping used for the PHY header and payload. The modulation and spreading functions for the PSSS mode PHY is shown in Figure 4–8.

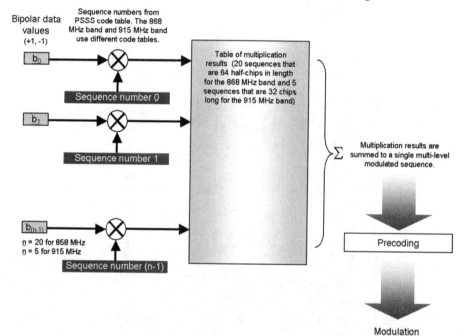

Figure 4–7: Symbol-to-chip mapping for PSSS mode PHY

RADIO CHARACTERISTICS

IEEE Std 802.15.4 radio specification was designed to allow the implementation of low-cost digital integrated circuit designs. Most of the technical radio specification can be considered relaxed with respect to other radio technologies. The following paragraphs present some of the characteristics of the IEEE Std 802.15.4 radio.

Power Output

IEEE Std 802.15.4 provides for a wide transmitter output power range, but the device must be capable of transmitting –3 dBm. The upper limit of power output

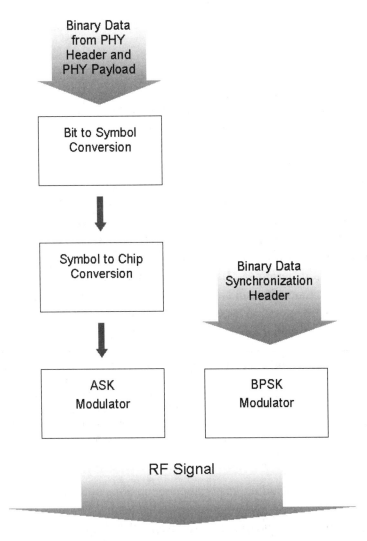

Figure 4–8:   868/915 MHz PSSS PHY modulation and spreading

is designated by the regulatory agency associated with the local use. For example, in the United States, some services using DSSS in the 2.4 GHz band are allowed up to 1 W of transmitter power [22]; however, in Europe the limit is 100 mW in the same band [6].

Sensitivity

IEEE Std 802.15.4 specifies that the receiver must be capable of correctly decoding a signal with an input power of –85 dBm or less in the 2.4 GHz band as well as the optional O-QPSK and PSSS PHYs in the 868/915 MHz band. In the lower frequency bands where BPSK is used, the receiver must be capable of correctly decoding a signal with an input power of –92 dBm or less. Better sensitivity is not prohibited.

Range

In free-space, the path loss between a transmitter and a receiver is dependent solely on the distance between them, and is independent of the frequency used. There is therefore no inherent difference in range between the IEEE Std 802.15.4 frequency bands. The type of antenna used, however, can give the illusion of a frequency-dependent path loss: A constant-gain antenna, such as a half-wave dipole, has an effective aperture (area) that decreases as the frequency of operation increases. (The dipole gets physically smaller as the frequency increases, leading to a reduction of its effective area.) At higher frequencies, the dipole therefore intercepts less of the transmitted radiation and produces less received signal at its terminals. Users of a communication link employing dipole antennas on both transmitter and receiver would therefore notice a range reduction as the frequency of operation increased, although the effect was due to the antennas employed, rather than an increase in path loss.

 Antennas that have a constant effective aperture exist; such antennas have gain that increases as the frequency of operation increases. An example of such an antenna is the parabolic dish antenna. Users of a communication link employing parabolic dish antennas of fixed dimensions on both transmitter and receiver would notice a range extension as the frequency of operation was increased—just the opposite effect of that seen by the users employing dipole antennas.

The previous notwithstanding, many implementations of the IEEE Std 802.15.4 are expected to employ constant-gain (dipole) antennas. For these implementations, in free space the ratio of the power available at the terminals of a receiving antenna to the power supplied to the terminals of a transmitting antenna is given by the Friis model:

$$\frac{P_R}{P_T} = G_T G_R \left(\frac{\lambda}{4\pi d}\right)^2 = G_T G_R \left(\frac{c}{4\pi f d}\right)^2$$

where P_R and P_T are the power values at the receiving antenna and transmitting antenna, respectively, (in watts); G_R and G_T are the power gains of the receiving antenna and transmitting antenna, respectively; λ is the wavelength (in meters); f is the frequency (in Hertz); d is the distance (in meters), and c is the speed of light (in meters/second).

Continuing under the assumption of constant-gain antennas, this same equation can be expressed as a basic transmit-antenna-terminal-to-receive-antenna-terminal loss L_B in decibel form, with the appropriate substitutions for the constants as:

$$L_B(dB) = 32.44 + 20\log_{10} f_{MHz} + 20\log_{10} d_{km}$$

in which the loss L_B includes contributions from both the antenna effective area (which decreases with frequency), and the path loss itself (which is constant with frequency). Here, since we assume constant-gain antennas, we have also normalized G_R and G_T to 1. For further simplification, we treat the 868 and 915 MHz bands as approximately 1 GHz and the 2.4 GHz band as approximately 2 GHz so that the losses can be expressed as:

$$L_{B\,@\,1GHz}(dB) \approx 92 + 20\log_{10} d_{km}$$

$$L_{B\,@\,2GHz}(dB) \approx 98 + 20\log_{10} d_{km}$$

Because lower frequency bands must exhibit a sensitivity of at least –92 dBm, for a transmitter rated at 0 dBm, and path loss of 92 dB, the maximum free space range is approximately 1 km ($\log_{10}[1]=0$).

For the higher 2.4 GHz band, the receiver must exhibit a sensitivity of –85 dBm; for a transmitter rated at 0 dBm, similar calculations will show a maximum free space range of approximately 220 meters. Note again that the range difference is due entirely to the type of antenna used.

Notice — Beware that the previous derivations are using the free-space radio propagation model. These calculations are ideal; make sure you finish reading this section, where a real-life model is used.

These ranges are for free space, using perfectly matched, unity-gain antennas, without interference, and represent the maximum theoretical distances. The free-space model does not consider several environment parameters that impair the RF propagation, including wave reflection, diffraction, and scattering. A commonly used model that approximates the propagation behavior in RF channels in real environments is the log-normal shadowing model. Figure 4–9 illustrates a typical indoor propagation effect and the variance in range obtained due to the effect of shadowing (transmitting a message with an output power of 0 dBm). The graphs show the received isotropic power vs. distance at 915 MHz and 2.4 GHz, respectively. As it can be seen, the distance decreases with increasing frequency, indicating the assumption of constant-gain antennas in this analysis. Similarly, the variability shown is a function of the characteristic of the wireless channel. The path loss coefficient in indoor environments can vary between $n = 2$ and $n = 4$ typically. Furthermore, the indoor variance in the received power ranges from $\sigma = 3$ to $\sigma = 11$.

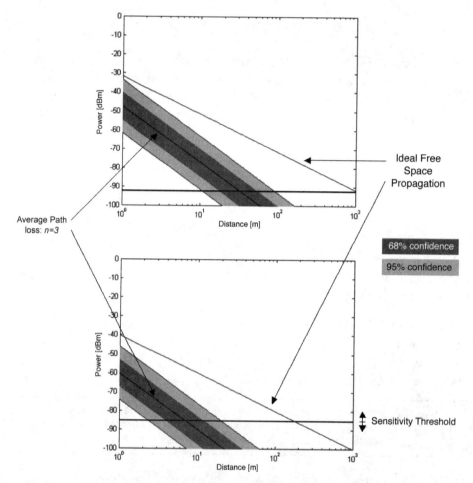

Figure 4–9:   Typical indoor range (0 dBm transmission) with normal distribution ($\sigma = 7$dB): (a) 915 MHz, (b) 2.4 GHz

Receiver Selectivity

The DSSS signal is, by definition, a wideband variation of the modulated data signal. The DSSS spreading provides the communications benefits already discussed, and it provides a relaxed RF selectivity requirement. In addition, IEEE Std 802.15.4 channels are widely spaced (5 MHz in the 2.4 GHz band) relative to their own bandwidth (3 MHz null-to-null; about 1.5 MHz noise bandwidth in the 2.4 GHz band). With the 0 dB adjacent channel requirement, as well as the relaxed requirement for channels further away, little selectivity is needed by the implementer to achieve the requirements.

Channel Selectivity and Blocking

IEEE Std 802.15.4 specifies an adjacent channel rejection for service in the 902–928 MHz band, where 10 channels are allowed, and in the 2.4 GHz band, where 16 channels are allowed. In the 868 MHz band only one channel exists and an adjacent channel specification is not meaningful. For the 902–928 MHz and 2.4 GHz bands, the receiver must reject an interfering adjacent channel signal that is at the same level (0dB difference) as a simultaneous on-channel signal. One interfering signal at a time is specified at that level.

In addition, IEEE Std 802.15.4 also specifies an "alternate" channel rejection for service in the 902–928 MHz and 2.4 GHz bands. The alternate channel is the next nearest channel to the adjacent channel, or two channels away from the channel of operation. The receiver must reject an interfering alternate channel signal that is at a level 30 dB higher than a simultaneous on-channel signal. One interfering signal at a time is specified at that level.

 The specification of these interference levels ensures reliable communication in the presence of multiple colocated WPANs, each on a different channel. The specification of a single interference signal reflects the relatively low traffic envisioned for the service.

To ensure that the receiver does not behave badly in other strong signal conditions, the specification provides a maximum acceptable input level that must not cause an excessive error rate. The IEEE Std 802.15.4 receiver must withstand a signal input of at least –20 dBm without unacceptable error rates.

There is no intermodulation specification in any of the IEEE Std 802.15.4 PHYs.

PHY SERVICES

The PHY layer provides an interface between the physical radio channel and the MAC sublayer, through the use of two services. These services are the PHY data service and the PHY management service (called the *PHY layer management entity* or PLME) and are accessed by the PHY layer data service access point (PD-SAP) and the PHY layer management entity service access point (PLME-SAP), respectively.

The services of the PHY layer are the capabilities it offers to the MAC sublayer.

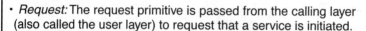

There are four types of service primitives. They are as follows:

• *Request:* The request primitive is passed from the calling layer (also called the user layer) to request that a service is initiated.

• *Indication:* The indication primitive is passed from the service layer to the user layer to indicate an internal event of significance. This event may be logically related to a remote service request, or it may be caused by a service layer internal event.

• *Response:* The response primitive is passed from the user layer to the service layer to complete a procedure previously invoked by an indication primitive.

• *Confirm:* The confirm primitive is passed from the service layer to the user layer to convey the results of one or more associated previous service requests.

PHY Data Services

The PHY data service provides three primitives to the MAC sublayer: PD-DATA.request, PD-DATA.confirm, and PD-DATA.indication, as shown in Figure 4–10. The data service does require a response from the peer MAC and so the response primitive is not required.

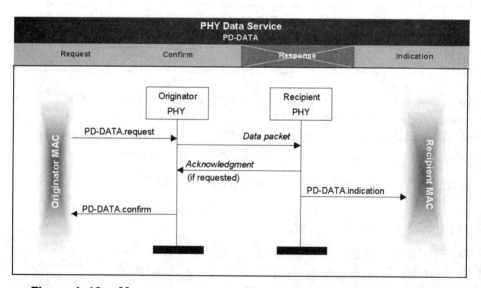

Figure 4–10: Message sequence diagram for data packet exchange mechanism

PHY Management Services

The PHY management service provides support for commands to control communication settings and the radio control functionality. The PLME primitives are summarized in Figure 4–11. The following paragraphs provide an overview of these primitives.

Primitive	Category	Description	Request	Confirm	Response	Indication
GET	Communication Settings	PHY PAN information base management	X	X		
SET			X	X		
SET-TRX-STATE	Radio Control	Enables/Disables radio system	X	X		
CCA	RF Energy Sensing	RF energy sensing: Clear Channel Assesment Energy Detection	X	X		
ED			X	X		

Figure 4–11:   PHY management service primitives

PHY PAN Information Base Management Primitives

The PHY PAN Information Base (PIB) contains configurable attributes to manage the PHY layer. These attributes can be read or written by the PLME-GET and PLME-SET primitives.

Figure 4–12 shows the message sequence diagrams for the procedure of reading or writing the PIB attributes.

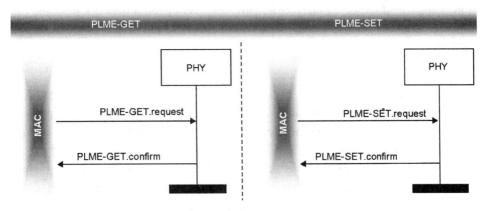

Figure 4–12:   Message sequence diagram for PHY PIB reading and writing mechanism

ENABLING AND DISABLING THE PHY

The radio transmitter and receiver can be enabled or disabled by means of the PLME-SET-TRX-STATE primitive. The purpose of this primitive is to control the radio transceiver and enable lower power consumption. Figure 4–13 shows the message sequence diagram for this primitive.

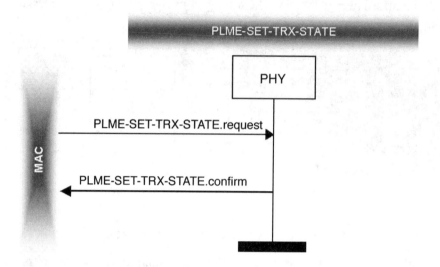

Figure 4–13:   Message sequence diagram for transceiver enabling mechanism

CLEAR CHANNEL ASSESSMENT

Before the transmission of packets in a non–beacon-enabled network or in the contention access period of a beacon-enabled network, the MAC instructs the PHY to perform a clear channel assessment (CCA) before sending data and MAC command frames.

When the clear channel assessment is requested, the PHY layer enables the receiver, performs a CCA measurement, and then disables the receiver. Once the clear channel assessment measurement is completed, the PHY layer issues an PLME-CCA.confirm indicating if the channel is busy or not. Figure 4–14 illustrates the CCA mechanism using the PLME-CCA primitives. The figure also shows the message sequence chart for the clear channel assessment mechanism.

ENERGY DETECTION

The PLME-ED primitive allows a device to perform RF energy detection in the actual channel where it is operating. The measurement performed is similar to the one carried out in the PLME-CCA primitive, but with higher resolution, it returns an energy level that ranges from 0 to 255. The use of this primitive can enhance the functionality of the network layers. Figure 4–15 shows the message sequence diagram for the energy detection process.

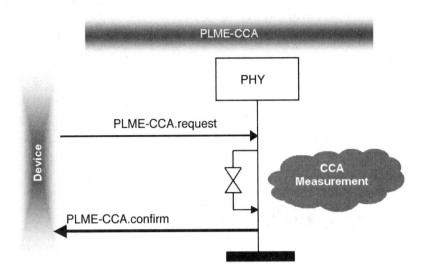

Figure 4–14: Message sequence diagram for the clear channel assessment mechanism

PHY PACKET STRUCTURE

The PHY Protocol Data Unit (PPDU) is the packet data structure at the PHY protocol level that modulates the wireless transmitter. The PPDU encapsulates all data structures from higher levels of protocol. The PPDU consists of three components: first, a synchronization header; second, a PHY header; and third, a variable length payload containing the PHY layer service data unit.

PPDU synchronization header — The PPDU synchronization header consists of two fields, a preamble and a start-of-frame delimiter. For all but the optional PSSS physical layer, the preamble consists of 32 bits, all set to binary zero (recall that a binary zero gets encoded in a chip pattern). For the optional PSSS mode the preamble is 40 bits long when operating at the 868 MHz band and is 30 bits long when operating at the 915MHz band. In both cases the preamble consists of the first sequence of the respective PSSS code table. The preamble field allows a receiver a sufficient number of bits to achieve chip and bit synchronization. The start-of-frame delimiter consists of the 8-bit pattern "0xe6" (11100101) and allows the receiver to establish the beginning of the packet in the stream of bits.

Note that the start-of-frame delimiter for the optional PSSS mode PHY differs from what is used by the other physical layer options. The start-of-frame delimiter for the PSSS PHY is the inverted first sequence in the respective PSSS code table.

PHY header — The PHY header is a single 8-bit field with the MSB reserved and the remaining low-order bits used to designate frame length information. Packet lengths of 0 to 4 and 6 to 8 bytes are reserved. Packets of length 5 bytes are MPDU acknowledgment packets, and packets with 9 or more bytes are MPDU payloads for the MAC protocol layer service.

PHY payload — The PHY payload is composed of only one field called the *physical layer service data unit* (PSDU). The PSDU is variable length in nature and carries the data payload of the PPDU. All packets carry an MPDU payload for the MAC layer.

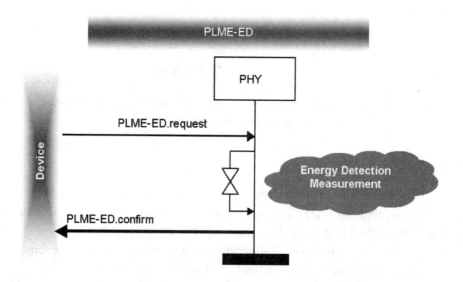

Figure 4–15:   Message sequence diagram for the energy detection mechanism

Figure 4–16 shows the structure of the PPDU.

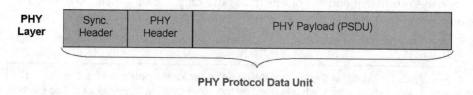

Figure 4–16:   PPDU structure

Chapter 5   Medium Access Control Sublayer

... arbitrating channel access and more

The MAC sublayer, together with the Logical Link Control (LLC) sublayer, comprises the data link layer (also called *layer 2*) in the ISO/OSI model [26]. The MAC layer provides access control to a shared channel and can offer a reliable data delivery. In the case of WPANs, optimal use of the wireless media is desirable, because they operate in restricted unlicensed bands shared by several other standard and non-standard wireless technologies, including WLANs. IEEE Std 802.15.4 uses a carrier sense multiple access with collision avoidance (CSMA-CA) algorithm, which requires listening to the channel before transmitting to avoid collisions with other on-going transmissions (a kind of wireless etiquette).

The IEEE Std 802.15.4 MAC sublayer has several functions, such as the generation of acknowledgment frames, association, disassociation, security control, beacon generation and optional guaranteed time slot management, which would be used in a star network topology. The IEEE Std 802.15.4 MAC was designed to allow the implementation of a very thin (simple) protocol stack. This facilitates the quick development of applications and has a direct impact on improving power consumption—the secret is simplicity!

The set of standards from IEEE 802 differ at the physical layer and MAC sublayer, but share a common interface at the data link layer. This is achieved through the standardized IEEE 802.2™ LLC. This standard LLC interface was defined prior to the introduction of wireless networking standards, i.e., WLAN and WPAN. For this reason, IEEE Std 802.15.4 defines the Service Specific Convergence sublayer, which makes the appropriate interface between the MAC and IEEE Std 802.2, while allowing the definition of other LLCs more appropriate for wireless media.

The IEEE 802.15.4 MAC sublayer definition contains enhanced functionality normally located in the LLC, making it suitable to be interfaced with the network layer directly, allowing the simple implementation of wireless devices.

The IEEE Std 802.15.4 MAC provides support for two types of wireless network topologies: star topology and peer-to-peer. The management of these types of networks is done in the network layer and is beyond the scope of IEEE Std 802.15.4. In this regard, the MAC only performs functions required by prospective network or other higher layers. Several application scenarios, mostly in the home networking segment, drove the vision for star networks, while for industrial and commercial applications the interest is more in peer-to-peer networking. The peer-to-peer topology enables the creation of larger ad hoc, self-organizing wireless networks (principally multihop/ mesh configurations).

All devices, regardless of topology, participate in the network using their unique 64-bit IEEE addresses; this address can be substituted for a short 16-bit address allocated by the PAN coordinator, thus ensuring that it is unique within its network; this process is managed by the association procedures explained in this chapter.

> **IEEE 802.15.4**
> **2006**
> In general, all MAC sublayer changes introduced in the 2006 revision are backward compatible with the original version. However, there are three exceptions. MAC Frames using security processing operations, frames using the channel page field, or frames with MAC payloads larger than 102 bytes are not backward compatible.

The 2006 revision of the original standard introduced some new features and modifications to the MAC sublayer. With the exception of security all these new features are backward compatible. In IEEE Std 802.15.4-2006, the security processing has been simplified and streamlined, reducing the security overhead. Other MAC changes include:

- Added a subfield for indicating the frame version.
- Added features facilitating the synchronization of network devices (mechanism to be implemented in higher layer).
- Made use of guaranteed time slots optional.
- Updated primitives to enable using the newly introduced channel pages as a result of the new PHY mode options.
- Allow scheduling the beacon start time to allow staggering of superframes.
- Simplified broadcasting in beacon enabled networks.
- Added features to allow reducing the association time in non-beacon-enabled networks.

Star Topology

In the star topology, communication is controlled by a single PAN coordinator that operates as a network master, sending beacons for device synchronization (including superframe control) and maintaining association management. In this topology, the network devices communicate only with the PAN coordinator, as shown in Figure 5–1.

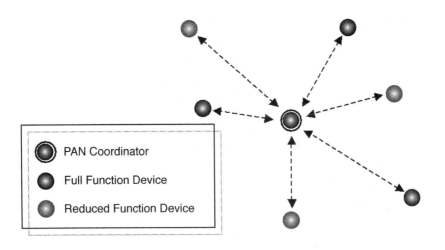

Figure 5–1: Star network topology

Any full-function device (FFD) may establish its own network by becoming a PAN coordinator. Each star network operates independently from any neighboring networks. During the formation of a new star network, the PAN coordinator must choose a network identifier, called a *PAN ID*, that is not used by any other network in its vicinity. This is accomplished by scanning all available or chosen channels for existing networks and then selecting a PAN ID that differs from the ones found. After this procedure is performed, the PAN coordinator can start sending beacons in regular intervals and permitting network devices, requesting to joins its network, to associate.

Network devices programmed to join a star network must first scan for available networks within their radio range by listening for beacons sent by a PAN coordinator. Upon completion of the scan, the higher layer of the network device may select to join one of the networks that were found by sending an association request to the PAN coordinator. The PAN coordinator in turn makes the decision allowing the network device to associate or not.

 Star networks also support a non–beacon-enabled mode. In this case, the PAN coordinator uses beacons for association purposes only. Network device synchronization for data exchange is achieved by polling the PAN coordinator for data on a periodic basis.

Peer-to-Peer Topology

The peer-to-peer topology allows any FFD to communicate with any other FFD within its range, and have messages relayed to FFDs outside its range, via multi-hop routing. This topology enables the formation of more complex, larger networks, including ad hoc, self-organizing, and self-healing structures. IEEE Std 802.15.4 does not specify the details of any of these networks; it only defines the MAC functionality to enable such capabilities.

RFDs may be employed in a peer-to-peer network, but only as peripheral devices, since they lack the capability to relay packets. As a consequence, a sufficient number of FFDs must be present to form the network. Figure 5–2 shows a typical peer-to-peer network.

Peer-to-peer communications requires additional device memory due to the size of the routing tables employed in the higher layer.

A specific type of a peer-to-peer network is the cluster-tree network, consisting of a variable number of coordinators that can serve as a cluster manager or "cluster head" within a cluster-tree structure but also as a router for relaying messages within the network. The main advantage of a multiclustered wireless network is its hierarchy, which can be used to greatly simplify the routing algorithms used to relay messages across the network. Figure 5–3 illustrates an example of a typical cluster-tree network; the figure shows network devices, cluster-heads, and the PAN coordinator that operates as the director of the operation for the entire network.

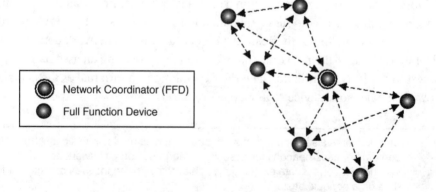

Figure 5–2:   Peer-to-peer network topology

Similar to the procedure for starting a star network, a new peer-to-peer network is formed by an FFD establishing itself as a PAN coordinator and selecting a PAN ID that differs from other networks within its vicinity. In most cases a peer-to-peer network will be non-beacon enabled, but it may operate in a beacon-enabled mode. Network devices programmed to join a peer-to-peer network must scan for available networks within its vicinity by discovering a nearby PAN coordinator or routing capable devices that act as proxies for PAN coordinators. The IEEE 802.15.4 standard refers to routing capable devices as coordinators. Upon completion of the scan, the higher layer of the network device may select to join one of the networks that were found by sending an association request to the PAN coordinator or alternatively to one of the closest coordinators. The coordinator in turn makes the decision allowing the network device to associate or not. If the network device is an FFD, it also becomes a coordinator once joined to a network and provides routing services to other devices in the network or trying to join the network.

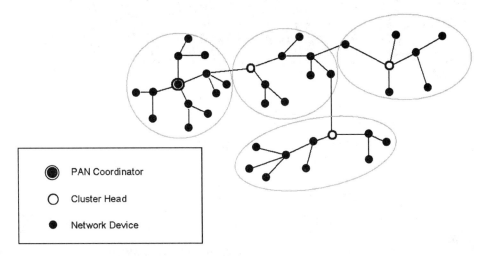

Figure 5–3:   Cluster-tree network topology

SUPERFRAME STRUCTURE

IEEE Std 802.15.4 allows the implementation of an optional superframe structure. The superframe is managed by the PAN coordinator and is bounded by beacon messages sent by the coordinator at programmable regular intervals, called the *beacon interval*. The structure of the superframe can be configured specifically to address the needs of various applications anywhere from small low-latency star networks to large, long-latency, multihop networks.

Each beacon contains information that will helps network devices synchronize to the network; this information includes the network identifier, beacon periodicity, and superframe structure. A superframe is divided into 16 contiguous time slots; the first time slot starts at the beginning of a beacon frame.

Network devices that need to communicate with the PAN coordinator must attempt to do it in the time between two successive beacons. This period of time is called the *contention access period* (CAP). To communicate with the PAN coordinator, each network device needs to access the channel using slotted CSMA-CA. Figure 5–4 shows the generic structure of a superframe.

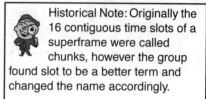

Historical Note: Originally the 16 contiguous time slots of a superframe were called chunks, however the group found slot to be a better term and changed the name accordingly.

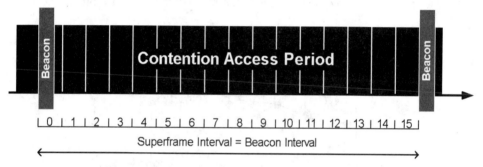

Figure 5–4: Generic superframe structure

Each GTS is formed by an integer multiple of time slots. Each time slot is equal to 1/16 of the time between the start of two successive beacons.

By request, the PAN coordinator can optionally assign dedicated portions of the superframe to a specific network device. These segments of time are called *guaranteed time slots* (GTSs). This capability supports applications with a particular bandwidth requirement or that need lower communications latency and is used in star networks. GTSs are all grouped toward the end of the superframe right before the next beacon as illustrated in Figure 5–5. The length of time covering all GTSs is defined as the contention-free period (CFP). Though the CFP may take up a significant portion of the superframe, the IEEE 802.15.4 standard does require that a minimum amount of CAP space must remain (440 symbols) to allow other devices, not using the CFP, to access to the channel. The only exceptions to this rule is temporary reductions needed to accommodate additional beacon lengths for GTS maintenance.

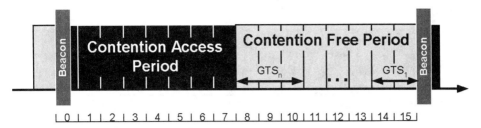

Figure 5–5: Superframe structure with GTS

The assignment of the GTSs is entirely at the discretion of the PAN coordinator, but the use of contention-free access is useful for applications requiring low latency and dedicated bandwidth allocation.

- *Low latency:* For those applications that require minimum end-to-end delay for prioritized messages, slot assignment facilitates preemptive, out-of-order queue management in the protocols to minimize that delay. Examples might include alarm conditions relaying, or QoS differentiation based on application software.

- *Bandwidth allocation:* For those applications that may be less sensitive to delay, but generate a known data traffic rate, the facility to allocate bandwidth to each service permits the protocol to manage queue lengths at each network device. Traffic management is usually associated with network protocols above the MAC layer, but the capability must be established at the lower levels covered by IEEE Std 802.15.4 to be available to higher levels.

GTSs make higher level protocol services possible, but for those networks that do not implement higher levels of service, the use of GTSs is optional.

For applications with relaxed latency requirements, the superframe may also be split into an active and an inactive portion with the 16 contiguous slots occupying only the active portion of the superframe. An inactive portion of the superframe is not used for communication between devices. Figure 5–6 shows an example where the active and inactive portion are of equal length, however this is not a requirement and they may be of different proportions. The length of the active portion of the superframe is specified by the superframe interval, while the frequency of beacons is specified by the beacon interval. This for instance, allows networks implemented with a battery-powered coordinator to reduce the communication duty cycle and increase the battery life of the coordinator.

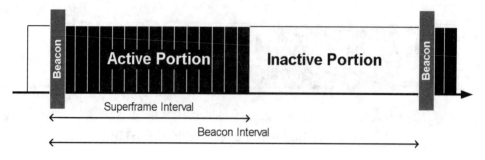

Figure 5–6:   Superframe structure with active and inactive portions

IEEE 802.15.4
2006 A new feature of the 802.15.4 MAC sublayer allows a coordinator to schedule the start of the beacon of a superframe in respect to its coordinator's beacon. This allows multihop networks to use the superframe structure enabling OEMs to create large battery-powered networks.

Another use of this feature is for applications requiring larger beacon-enabled, multihop networks. This feature allows staggering of multiple superframes or without interfering with another. In this case a coordinator (a device providing coordination services to other devices) aligns the start of its active period to fit within the inactive period of its parent's superframe. The parent node may be the PAN coordinator or even another coordinator as see in Figure 5–7.

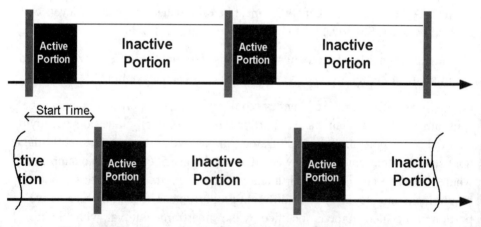

Figure 5–7:   Superframe structure in multihop networks

MAC DATA TRANSFER MODEL

The shared nature of the radio-frequency spectrum presents several challenges to the wireless communication system designer. The IEEE Std 802.15.4 MAC was designed to take these challenges into account and provide measures to improve communications reliability by offering a fully acknowledged protocol using CSMA-CA and a frame integrity check.

The data transfer model of IEEE Std 802.15.4 depends on the network topology. In star networks the communication exchange always occurs between a PAN coordinator and a network device, but in the peer-to-peer mode, a device may communicate with its coordinator or any other in its vicinity.

Star networks can have two types of data transfer mechanisms depending on whether the PAN coordinator is beacon-enabled or not. These are data transfer to a PAN coordinator and the data transfer from a coordinator. A peer-to-peer network supports in addition also the peer-to-peer data transfer.

Data Transfer to a Coordinator

In a beacon-enabled network, a device with data for the coordinator needs to be synchronized to the beacons, sent periodically by the coordinator. If the device is part of a star network and has a GTS assigned, it waits for the appropriate point within the superframe to transmit its data frame without using slotted CSMA. Otherwise the device transmits its data frame in the contention access period of the superframe, in accordance with the slotted CSMA-CA procedure. After receiving the data frame, the coordinator may return an acknowledgment to the network device, if requested; at which point the data transfer is completed. The message sequence diagram for this process is shown in Figure 5–8. The IEEE Std 802.15.4 refers to this procedure as direct data transfer.

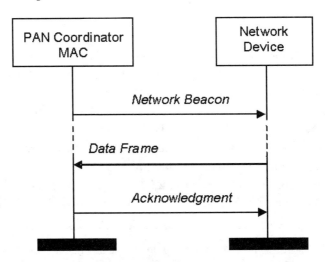

**Figure 5–8: Data transfer to a coordinator—
beacon-enabled network**

A device with data to send to its coordinator in a non-beacon-enabled network simply checks the channel availability using CSMA and if the channel is idle sends the message to the coordinator. If an acknowledgement is requested, the coordinator confirms the successful receipt of the data frame by returning an acknowledgement to the device. This procedure is outline in Figure 5–9

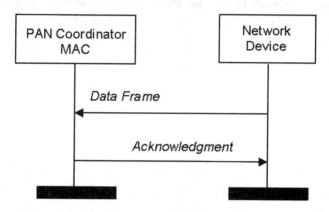

Figure 5–9:   Data transfer to a coordinator—non-beacon-enabled network

Data Transfer from a Coordinator

When a coordinator has data to transmit to a device it does not transmit the message immediately but instead appends the device's short address in a special field, called the *pending address list,* in its beacon. This indicates to the device that data is pending for it at the coordinator. Once the appropriate device receives the beacon and detects that data is pending, it sends a data request MAC command frame to the coordinator in the CAP of the superframe. Upon receipt of the command frame, the coordinator responds with an acknowledgment frame followed by the pending data frame. The transaction is completed by an acknowledgment sent from the device to the coordinator indicating the successful receipt of the data frame. If additional messages are pending for the same device the coordinator posts an indication in the beacon of the following superframe. The IEEE Std 802.15.4 refers to this procedure as indirect message transfer, the message sequence diagram of this process is summarized in Figure 5–10.

 The IEEE Std 802.15.4 design team decided not to implement a combined acknowledge/data frame in order to simplify implementation.

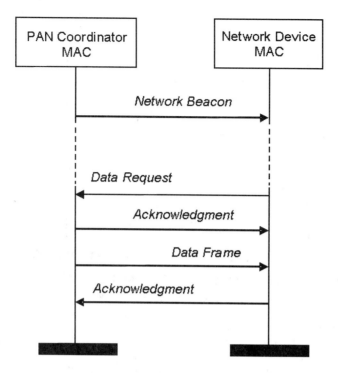

**Figure 5–10:   Data transfer from a coordinator—
beacon-enabled network**

Indirect data transfer is also used by a coordinator to send data to a device in a non-beacon-enabled network. However, since there is no mechanism for the coordinator to indicate to the device that a message is pending, the device needs to frequently poll the coordinator for pending messages. The IEEE Std 802.15.4 MAC sublayer has a mechanism allowing the higher layer to poll for messages. Figure 5–11 shows the procedure of the indirect data transfer in a non-beacon-enabled network. When the higher layer of a device instructs the MAC to poll for pending messages, the MAC sends a data request MAC command frame to the coordinator, which confirms the successful receipt of the command with an acknowledgement. This is followed by the pending data frame sent from the coordinator to the device. The transaction is completed by an acknowledgement sent from the device to the coordinator confirming the successful receipt of the data message.

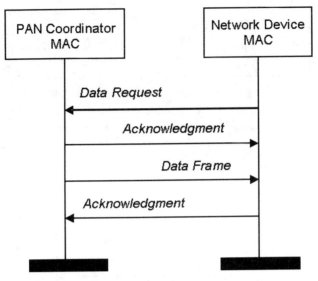

Figure 5–11: Star network data transfer from PAN coordinator—non-beacon-enabled network

Peer-to-Peer Data Transfer

For peer-to-peer topologies the data transfer strategy is governed by the specific network layer managing the wireless network. A given network device may stay in reception mode, scanning the radio-frequency channel for on-going communications, or may send periodic beacons to achieve synchronization with other potential listening devices.

MAC SERVICES

The MAC sublayer provides two services to the higher layers. These services are the MAC data service and the MAC management service (called the *MAC sublayer Management Entity* or MLME). They are accessed by the MAC common part sublayer service access point (MCPS-SAP) and the MAC management service access point (MLME-SAP), respectively. For each of these services, IEEE Std 802.15.4 defines a set of protocol primitives that enables the full functionality of LR-WPAN devices. Figure 5–12 shows the detailed IEEE Std 802.15.4 protocol stack architecture and the virtual link between MAC peer entities.

The services of a layer are the capabilities it offers to the next higher layer (or sublayer). The upper layers build their functionality based on these lower layer services. The MAC services allow the transport of protocol data units between peer entities.

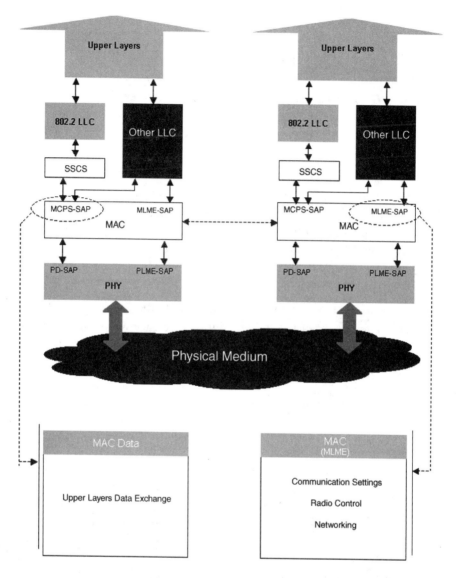

Figure 5–12:   IEEE Std 802.15.4 protocol stack architecture

Transmission Scenarios

The packet exchange between two transceivers is susceptible to errors due to the nature of the wireless medium. When one or more errors are present in any given frame, the message will not arrive at the intended MAC recipient because either the recipient PHY layer rejected the packet or the PHY could not synchronize (correlate) with the message itself (in this case, the PHY did not "hear" the message).

In any acknowledged data transfer there are three possible transmission scenarios, as follows:

- *Successful Data Transmission:* The originator device sends a message to a recipient MAC. The message is received with no errors detected and an acknowledgment message is sent back to the originator before a timeout occurs. A successful data transmission scenario is illustrated in Figure 5–13.

- *Lost Message Frame:* A message addressed to an intended MAC recipient never arrived at its destination. In this case, the originating device will declare a timeout and will make another attempt to send the message. After a number of attempts, the MAC of the originating device will send a notification to its upper layer indicating transmission failure. This transmission scenario is shown in Figure 5–14.

- *Lost Acknowledgment Frame:* The originator of the message transaction did not receive an acknowledgement frame. Similar to the previous case, after a number of attempts, the MAC of the originating device will send a notification to its upper layer indicating transmission failure. This transmission scenario is shown in Figure 5–15.

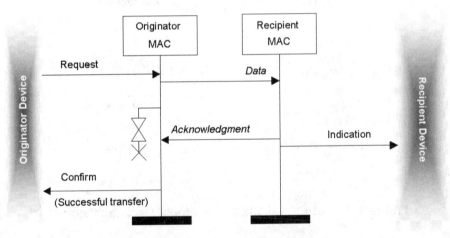

**Figure 5–13:   Transmission scenario —
successful data transfer**

MAC Data Service

The MAC data service provides three primitives for data transfer: MCPS-DATA.request, MCPS-DATA.confirm, and MCPS-DATA.indication, as shown in Figure 5–16. This figure shows the sequence diagram for a typical data

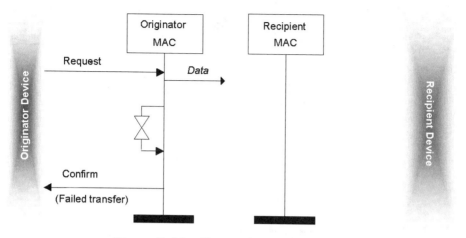

**Figure 5–14: Transmission scenario—
lost message frame**

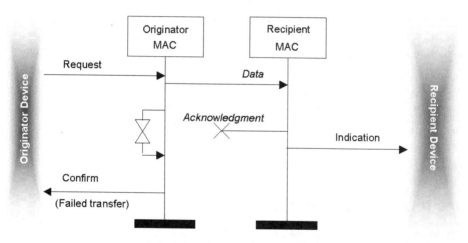

**Figure 5–15: Transmission scenario—
lost acknowledgment frame**

transfer between two devices. The data service does require a response from the peer MAC, and so the response primitive is not required.

The interface offered by the MAC data service has the capability of handling diverse addressing schemes that enable simple implementation of star and peer-to-peer network topologies. In this sense, the MAC enables the upper layers to control which address fields are present in the packet to be transmitted, allowing the implementation of diverse types of network layers, including multihop ad hoc

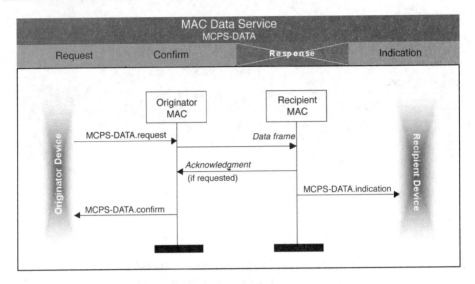

Figure 5–16:   Message sequence diagram for data exchange mechanism

networks. IEEE Std 802.15.4 uses a standard 64-bit IEEE address and short addresses allocated by the association mechanisms. In addition, the MAC data service supports two optional primitives that enable the upper layers to purge an MSDU from the MAC queue. These primitives are MCPS-PURGE.request and MCPS-PURGE.confirm and are used for the indirect data transfer. This can be used, for instance, when the information contained in a pending data frame expires before the recipient device gets a chance to retrieve it from the coordinator.

MAC Management Service

The MAC management service provides support for commands to control communication settings, the radio control, and networking functionality. The MLME primitives are summarized in Figure 5–17. The following paragraphs provide an overview of these primitives.

To enable the implementation of very low complexity devices, some of the MAC management primitives are optional (i.e., they do not need to be implemented in order for a product to be IEEE Std 802.15.4 compliant). These primitives are MLME-GTS, MLME-RX-ENABLE, and MLME-SYNC. While other primitives are optional for RFDs only. These primitives are MLME-ASSOCIATE.indication, MLME-ASSOCIATE.response, MLME-ORPHAN, and MLME-START.

Primitive	Category	Description	Request	Confirm	Response	Indication
GET	Communication Settings	MAC PAN information base management	X	X		
SET			X	X		
RESET			X	X		
RX-ENABLE	Radio Control	Enables/Disables radio system	X	X		
SCAN		Scan radio channels	X	X		
ASSOCIATE	Networking	Association control with a network coordinator	X	X	X	X
DISASSOCIATE			X	X		X
GTS		GTS Management	X	X		X
ORPHAN		Orphan device management			X	X
SYNC		Control of device syncronization with network coordinator	X			
SYNC-LOSS						X
START		Beacon Management	X	X		
BEACON-NOTIFY						X
POLL		Beaconless Syncronization	X	X		
COMM-STATUS		Communication Status				X

Figure 5–17: MAC management service primitives

MAC PAN Information Base Management Primitives

The MAC PAN Information base (PIB) contains configurable attributes to manage the MAC sublayer. These attributes can be read using the MLME-GET primitive and written using the MLME-SET primitive. In addition,

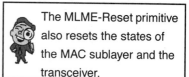 The MLME-Reset primitive also resets the states of the MAC sublayer and the transceiver.

the attributes can be reset to their default values with the MLME-Reset primitive.

Figure 5–18 shows the message sequence diagrams for the procedure of reading or writing the PIB attributes.

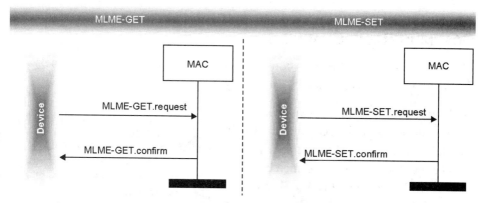

Figure 5–18: Message sequence diagram for reading and writing to the MAC PIB

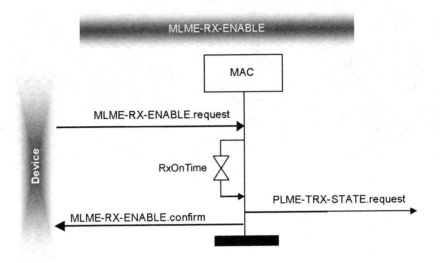

Figure 5–19: Message sequence diagram for enabling the receiver

ENABLING AND DISABLING THE RECEIVER

The radio receiver can be enabled or disabled by means of the MLME-RX-ENABLE primitive. The action can be set for immediate execution or scheduled for a later time. The purpose of scheduling is to enable the MAC to offer network synchronization capabilities to the upper layers.

This primitive allows the power consumption to be lowered in systems that use a network layer that takes advantage of this feature i.e., scheduling data exchange at determined instants, and minimizing power consumption. Figure 5–19 shows the message sequence diagram for this primitive.

SCANNING RADIO CHANNELS

MLME-SCAN allows the initiation of a scan over a list of available channels. There are four types of channel scans. They are as follows:

- *Energy Detection Scan:* This scan allows measuring the radio frequency energy in each of the logical channels specified. The measurement is performed by the MLME by issuing a PLME-ED.request primitive and is used by a PAN coordinator to find a suitable channel for starting a new network.

 Beacons are not only sent by PAN coordinators. FFDs associated with a PAN and configured as a coordinator can advertise their presence to other devices by sending beacons for either device synchronization or to facilitate device discovery.

- *Active Channel Scan:* This scan searches for PAN coordinators or coordinators in the radio sphere of influence of devices that are participating in beacon-enabled or non-beacon-enabled networks. For each logical channel, the device first sends a *Beacon Request* Command (see "MAC Command Frame" on page 98), which causes any PAN coordinator or coordinator to send a beacon. If the PAN coordinator or coordinator are part of a non-beacon-enabled network they will send the beacon using unslotted CSMA. If the PAN coordinator or coordinator are part of a beacon-enabled network, they will send the beacon at the next scheduled beacon interval.

- *Passive Channel Scan:* This scan is used to search for coordinators in the radio sphere of influence of devices, participating in a beacon-enabled network. *Passive* means that a channel is scanned just by passively listening without sending a *Beacon Request* Command.

- *Orphan Channel Scan:* Allows an orphaned device (a device that has lost connection with its coordinator) to perform a scan to locate a coordinator. This search is performed over the specified list of logical channels.

The general procedure of the radio scan is shown in the message sequence diagram in Figure 5–20. The scan time is the duration the MAC stays on a particular channel. This time period is based on the smallest superframe size and is calculated using a parameter passed with the primitive from the upper layer.

A device becomes an orphan when it determines that it has lost communication with its coordinator. Communication may be lost for a number of reasons including:

- Fading or interference on the existing channel

- A switch of the coordinator to another channel, due to degradation of the existing channel (e.g., interference)

- Motion of either the device or the coordinator, so that the two are now out of range of each other

ASSOCIATION AND DISASSOCIATION CONTROL

A device that is currently not associated to any network uses the channel scan procedure to find prospective candidate networks. If the device wishes to participate in a beacon enabled network it uses a passive scan. If it is trying to join a non-beacon-enabled network it uses an active scan. Following a successful active or passive channel scan the information collected from all the received beacons is passed on to the higher layer. The higher layer may then choose a candidate coordinator to join. If the device is trying to join a beacon enabled network it first synchronizes to the periodic beacons of the chosen coordinator. This is done by

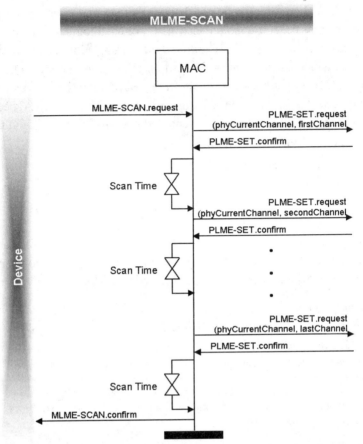

Figure 5–20: Message sequence diagram for the channel scanning mechanism

the use of the MLME-SYNC.request primitive (see "Synchronization Control" on page 90). Once synchronized, the higher layer of the device issues a MLME-ASSOCIATE.request primitive to the MAC sublayer. In a non-beacon-enabled network, the device does not need to synchronize, and the higher layer may issue the association request without synchronizing first.

Upon receipt of an association request primitive, the MAC of the device requesting association sends an *association request* command to the chosen coordinator. When the command has been received by the coordinator, it returns an acknowledgement to the device. Note that acknowledgements are optional for data frames, but are required for this procedure. The acknowledgment of this message does not imply that the association was accepted; it is simply a confirmation that the command was received. After receiving the association request, the MAC of the PAN coordinator needs to determine if it has sufficient resources to allow another device in its network. The PAN coordinator needs to make this decision in the time specified in the *macResponseWaitTime* (a MAC constant parameter) or the requesting device will declare a timeout.

The details of the process for selecting a given PAN coordinator after a channel scan are not defined in the IEEE 802.15.4 standard. The upper layers of a particular implementation must define an association strategy.

Depending on the capabilities and the requirements of the application, the coordinator can accept or reject the association request by issuing an MLME-ASSOCIATE.response with the appropriate parameters. The association response primitive is sent from the coordinator to the device requesting association using indirect data transfer (see "Data Transfer from a Coordinator" on page 74). If coordinator is part of a beacon enabled network, it will indicate in its beacon that a message is pending for the device trying to join. If the device is trying to join a non-beacon-enabled network, it will wait for *macResponseWaitTime* before requesting a response from the coordinator.

Another important feature of the association process is the possibility to request a short 16-bit allocated address from the PAN coordinator. This enables better bandwidth utilization because it reduces the total length of the packet (16-bit as opposed to 64-bit addresses). If the associating device does not request a 16-bit address allocation, the device will participate in the network using its extended 64-bit unique address. Figure 5–22 illustrates the message sequence diagram of the association process.

A network device or the PAN coordinator can initiate the dissociation process by using the MLME-DISASSOCIATE primitives. The higher layer can request to disassociation by issuing the MLME-DISASSOCIATE.request primitive to the MAC sublayer. This causes the MAC sublayer to send a disassociation notification command frame using either the data transfer to a coordinator or data transfer from a coordinator depending on which device initiated the request. The message sequence diagram of this process is shown in Figure 5–21.

GUARANTEED TIME SLOT MANAGEMENT

As described previously, IEEE Std 802.15.4 supports the optional use of a superframe, which enables the implementation of guaranteed time

The support for GTS was made optional in the 2006 revision of the standard.

IEEE 802.15.4
2006

slots. The MLME-GTS primitives allow the allocation of new a GTS, deallocation

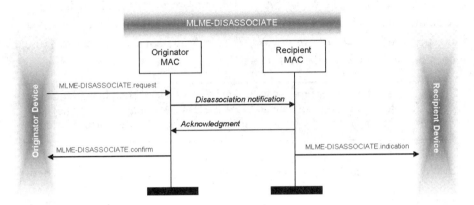

Figure 5–21: Message sequence diagram for the disassociation mechanism

of an existing GTS, or the reallocation of a GTS to eliminate time slot fragmentation.

A single GTS can extend over one or more superframe slots. The management of GTSs is performed only by the PAN coordinator, which controls how many of the 16 available time slots are assigned to the contention-free period (the rest are assigned to the contention access period). The PAN coordinator may allocate up to seven GTSs.

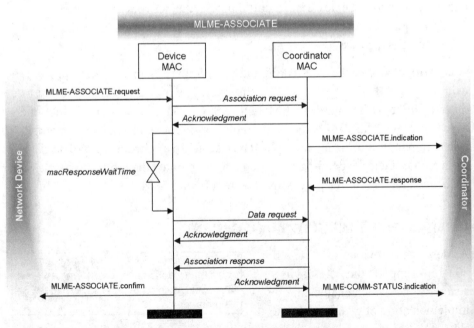

Figure 5–22: Message sequence diagram for association mechanism

The optional GTS capability allows the implementation of wireless links between a PAN coordinator and a network device with a guaranteed throughput (assuming a reliable link).

The PAN coordinator can potentially assign a single GTS covering the whole contention-free period, which in turn may occupy the whole superframe period, less the minimum length contention access period of 440 symbols.

The GTS allocation mechanism is initiated by the network device as indicated in the GTS message sequence diagram of Figure 5–23. The GTS deallocation mechanism can be initiated by the network device or by the PAN coordinator as indicated in the GTS message sequence diagram of Figure 5–24. On reception of a *GTS request* message, the PAN coordinator determines if space is available within the superframe structure based on the remaining length of the contention period and the number of time slots requested.

When the PAN coordinator is responding to a GTS request, it will generate a beacon with its GTS fields indicating the time slot allocated and the number of slots assigned; if there is no capacity in the superframe to allocate a new GTS of the desired characteristics, the PAN coordinator will deny the request. In this case, the GTS fields in the beacon will indicate the number of time slots remaining in the contention-free period.

 Transmissions from network devices in a GTS do not require the use of CSMA-CA.

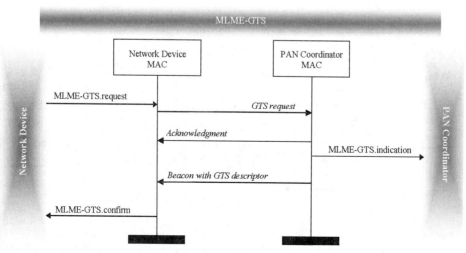

Figure 5–23: GTS allocation procedure

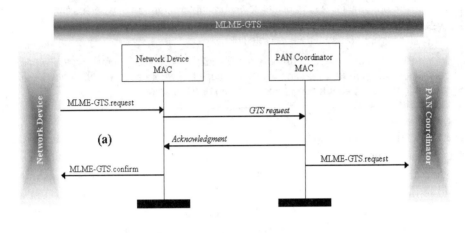

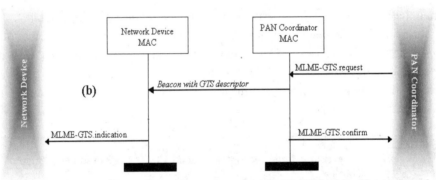

Figure 5–24: GTS deallocation procedure (a) initiated by the network device and (b) initiated by the PAN coordinator

When a network device generates a data request in a GTS-enabled network, the device will have to defer transmissions until the start of the assigned GTS. This transmission cannot take more time than the allocated time (number of allocated time slots).

The GTS message exchange mechanism is fully acknowledged. It concludes with the generation of an MLME-GTS.confirm primitive that includes all relevant information of the GTS allocation procedure. After being assigned a GTS, a network device will use its allocated time to communicate with the PAN coordinator, although the network device will also be able to communicate with the PAN coordinator using the contention access period.

 The allocated guaranteed time slots are directional, with data transfer either from the device to the PAN coordinator or from the PAN coordinator to the device.

A network device or the PAN coordinator can request the deallocation of an existing GTS. If the operation is initiated by the network device, its MLME should be instructed via an MLME-GTS.request with the appropriate parameters. Similarly, when the PAN coordinator initiates the deallocation of a given GTS, it should send a beacon with its GTS fields indicating that the GTS assigned is being deallocated. The complete deallocation process follows the same steps as the allocation mechanism as presented in Figure 5–24. After the deallocation of a given GTS, the network device can continue communicating with the PAN coordinator using the contention access period.

On a GTS-enabled network, the contention-free period of the superframe can become fragmented as a result of several allocations and deallocations. The PAN coordinator is in charge of removing gaps within the contention-free period in order to ensure optimal use of bandwidth.

If a device loses synchronization with the PAN coordinator, the GTS will be deallocated. After reestablishing a connection with the PAN coordinator (following a scan process), the device may request GTS allocation again.

ORPHAN DEVICE MANAGEMENT

In the event that a network device loses contact with its PAN coordinator, it will perform an orphan channel scan using the MLME-SCAN primitive. As part of the channel scan, the MAC of the network device will send *orphan notification* command messages in each of the available and specified channels. When the MAC of the PAN coordinator or a coordinator receives this notification, it generates an MLME-ORPHAN.indication, which causes the coordinator to verify if the network device was previously associated with its network. If the network device was associated to the coordinator, it will generate an MLME-ORPHAN.response. At this point the MAC of the coordinator will send a *coordinator realignment* command. In the event that the network device was not previously associated with the PAN coordinator, there will be no response from it. Figure 5–25 shows a message sequence diagram for the orphan notification mechanism.

A reduced function device is not required to implement the MLME-ORPHAN primitives.

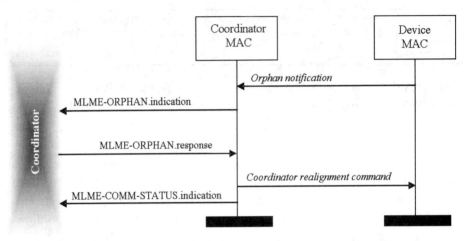

Figure 5–25:   Message sequence diagram for orphan notification mechanism

SYNCHRONIZATION CONTROL

Synchronization control with a coordinator is achieved with the use of the MLME-SYNC and MLME-SYNC-LOSS primitives. MLME-SYNC allows the network device to locate and track beacons in a beacon-enabled PAN. The search process is initiated with an MLME-SYNC.request. The search is done by activating the radio receiver and waiting, for a given amount of time, for a beacon frame to arrive from a coordinator.

In a beacon-enabled PAN, the search for beacons can be performed in one of two modes: one mode in which the MAC of the network device continuously tracks its coordinator's beacons; or another mode in which the MAC locates the beacon only once. In both cases, if a beacon is received indicating data pending for the receiving network device, it sends a data request command to the PAN coordinator. The message sequence chart for network device synchronization is shown in Figure 5–26. The MLME-SYNC.request is not used in non beacon-enabled PANs.

In the event of a loss of synchronization with the coordinator, the MAC generates an MLME-SYNC-LOSS.indication. Four possible situations can cause a loss of synchronization event. They are as follows:

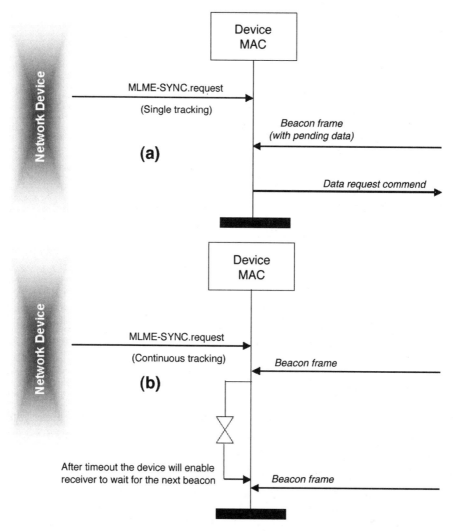

Figure 5–26:   Message sequence diagram for network device synchronization (a) single tracking and (b) continuous tracking

- *Beacon Lost:* After MLME-SYNC, either initially or during tracking, the beacon message has not been received. The message sequence chart for this scenario is shown in Figure 5–27.

- *Coordinator Lost:* Several attempts to communicate with the coordinator have failed.

- *PAN ID Conflict:* The network device detected a PAN identifier conflict, defined as two different PAN coordinators with the same ID being in range of the network device.

- *Realignment:* The network device received a coordinator realignment command message from the PAN coordinator.

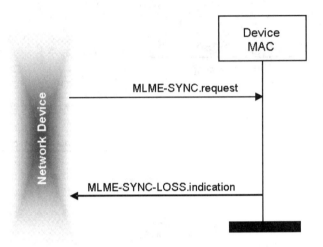

Figure 5–27:   Message sequence chart for loss of synchronization due to beacon lost scenario

BEACON MANAGEMENT

Beacon generation is initiated with the use of the MLME-START primitive. The parameters of this primitive allow configuring the device as a PAN coordinator or a coordinator, selecting a logical channel, establishing beacon periodicity, and setting up the superframe characteristics. After the MLME-START.request, the MAC will respond with an MLME-START.confirm as illustrated in Figure 5–28.

When a network device receives a beacon frame containing the PAN identifier of the network with which it is associated, the MAC will interpret its content. If the beacon frame contains one or more octets of payload (data), the MAC will issue an MLME-BEACON-NOTIFY.indication event. Figure 5–29 illustrates the beacon notification mechanism.

Full-function devices operating in a peer-to-peer network can generate beacon messages to neighboring devices. These beacons can act as "Hello" messages during network formation. If the MLME-START is issued with the *PANCoordinator* parameter set to FALSE, the device generating beacons is called a Coordinator.

In IEEE Std 802.15.4, only full-function devices can become coordinators and have the capability of generating beacons.

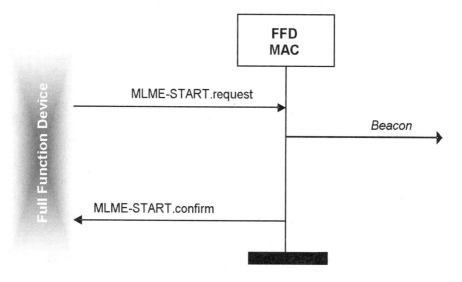

Figure 5–28:   Message sequence chart for procedure to start generating beacon transmissions

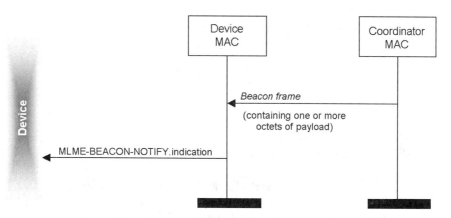

Figure 5–29:   Message sequence chart for beacon notification mechanism

BEACONLESS SYNCHRONIZATION

For non-beacon-enabled networks, a network device can use the MLME-POLL primitive to poll a coordinator for pending data at its discretion. When the MAC sublayer of a network device receives a MLME-POLL.request, it sends a *data request* command frame to its coordinator. The frame pending flag (FP) in the frame control field of the acknowledgement following the *data request* indicates

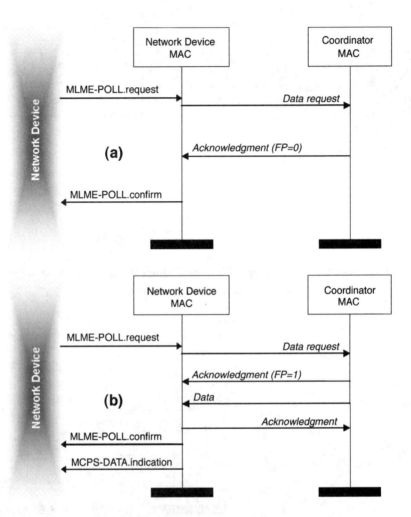

**Figure 5–30: Message sequence chart for polling the coordinator
(a) no data pending and (b) with data pending**

whether data is pending. After a MLME-POLL.request, the MAC will generate an MLME-POLL.confirm with the results of the polling procedure. Figure 5–30 shows the message sequence chart for this primitive.

If the network device received a beacon generated by its coordinator, the MAC will check its device address in the address list field from the beacon frame. If its address is listed, the MAC will generate a *data request* message to the coordinator, as indicated in the data transfer from the coordinator (beacon enabled).

COMMUNICATION STATUS

With this primitive, the MAC generates a variety of communication status messages to its upper layer. The MLME-COMM-STATUS.indication is invoked when any of the following conditions occur:

- After a transmission caused by a "response" primitive.
- After reception of a frame that did not pass the security procedure.

MAC FRAME STRUCTURE

The IEEE Std 802.15.4 MAC frame structure is designed in a way that reflects the protocol simplicity and flexibility while having the minimal elements to overcome the challenges of the wireless media. A MAC frame consists of three parts: header, variable length payload, and footer.

The MAC header contains a frame control field, a sequence number field, an addressing field, and optionally an auxiliary security header field. The frame control field specifies the type of frame, security usage, and the format and content of the address fields. The frame control field also indicates if an acknowledgment from the recipient of the frame is required. The sequence number field contains a number that is increasing with each transmitted frame. The address field contains the source or destination addresses as specified in the frame control field. If security is used (as indicated by a subfield in the frame control field) the frame may also include an auxiliary security header field. Its length may vary and depends on the specified security level.

 Depending on the type of MAC frame and the network topology in which a device is operating, a MAC frame can contain one or more of the following: the source (originator) PAN ID, source device address, destination PAN ID, and destination address. This feature enables the application to use diverse network topologies.

The MAC payload contains information specific to the type of transaction being handled by the MAC and can be logically divided in several fields for use by upper protocol layers.

Finally, the MAC footer consists of a 16-bit frame check sequence (FCS) based on the standardized ITU-T 16-bit cyclic redundancy check (CRC) algorithm (formally called CCITT 16-bit CRC). The general MAC frame structure is shown in Figure 5–31.

When the three components of the MAC frame are assembled into the PHY packet, it is called the *MAC protocol data unit* (MPDU).

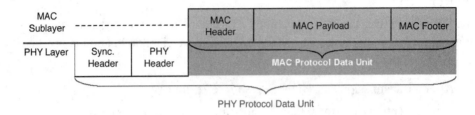

Figure 5–31: General MAC frame structure

IEEE Std 802.15.4 defines four types of MAC frames: Beacon, Data, Acknowledgment, and MAC command. These frames are described in the following paragraphs.

Beacon Frames

In a beacon-enabled network, a full-function device may transmit beacon frames. In a beacon frame, the address field contains the source PAN ID and the source device address. The MAC payload of a beacon frame is divided into four fields:

- *Superframe Specification Field:* Contains the parameters that specify the superframe structure (if any).
- *Pending Address Specification Field:* Contains the number and type of addresses listed in the Address List Field.
- *Address List Field:* Contains a list of device addresses with data available at the PAN coordinator.
- *Beacon Payload Field:* Optional field that may be used by higher layers. For example, it could be used to broadcast data to the devices participating in its network within its range of coverage.

The format of the beacon frame is shown in Figure 5–32.

MAC Header				MAC Payload				MAC Footer
Frame Control	Sequence Number	Addressing Fields	Auxiliary Security Header	Superframe Specification	GTS fields	Pending address field	Beacon Payload	FCS

Figure 5–32: Beacon frame format

Data Frames

Data frames are used by the MAC sublayer to transmit data. The address field will contain the PAN ID and device ID of the source and/or destination, as specified in the MCPS-DATA.request primitive. The format of the data frame is shown in Figure 5–33.

MAC Header				MAC Payload	MAC Footer
Frame Control	Sequence Number	Addressing Fields	Auxiliary Security Header	Data Payload	FCS

Figure 5–33:   Data frame format

Acknowledgment Frame

Acknowledgment frames are sent by the MAC sublayer to confirm successful frame reception to the originator of a message. The acknowledgment frame is generated only if the received message is requesting acknowledgment and the FCS (frame check sequence) is evaluated as good by the receiving device.

The acknowledgment frame does not contain the address field in the MAC header. Similarly, there is no MAC payload.

When a network device receives an acknowledge frame, it first verifies that it was expecting one and then matches the received

 The acknowledgment was designed to be very short, in order to minimize network traffic. Recall that when a message acknowledgment is required, there will be one acknowledgment per MAC frame.

frame sequence number with the one it is expecting; otherwise, the acknowledgment frame is discarded. Figure 5–34 shows the format of the acknowledgment frame.

MAC Header		MAC Payload	MAC Footer
Frame Control	Sequence Number		FCS

Figure 5–34:   Acknowledgment frame format

Table 5–1: MAC command frame types

Command Identifier	Command Type
1	Association Request
2	Association Response
3	Disassociation Notification
4	Data Request
5	PAN ID Conflict Notification
6	Orphan Notification
7	Beacon Request
8	Coordinator Realigment
9	GTS Request
10-255	Reserved

MAC Command Frame

The MAC command frame is originated by the MAC sublayer and is in charge of all the MAC control transfers for each MAC command type shown in Table 5–1.

The MAC payload has two fields, the MAC command type and the MAC command payload. The MAC command payload contains information specific to the type of command in use. Figure 5–35 shows the details of the MAC command format.

MAC Header				MAC Payload		MAC Footer
Frame Control	Sequence Number	Addressing Fields	Auxiliary Security Header	Command Type	MAC Command Payload	FCS

Figure 5–35: MAC command frame format

MAC FUNCTIONAL SCENARIOS

IEEE Std 802.15.4 contains a detailed explanation of the mechanisms that define the MAC functionality. The following paragraphs provide a brief overview of these mechanisms.

Accessing the Channel

In IEEE Std 802.15.4, the physical radio channel is accessed using CSMA-CA before attempting to send any frame. Exceptions to this are beacon frames, GTS transmissions, acknowledgment frames, and data frames following a data request command frame if the MAC sublayer can return the data frame within 12–32 symbols after the command frame.

 In beacon-enabled PANs, slotted CSMA-CA is used, whereas in non-beacon-enabled PANs, the basic unslotted CSMA-CA algorithm is implemented.

Starting and Maintaining PANs

Network devices attempting to participate in a network must proceed to locate it by scanning the RF channels in its list of available channels. As previously described, the MLME-SCAN primitive allows performing active and passive scanning for beacons. After a beacon of a suitable coordinator is found, the network device initiates the association process.

Devices that lose communications with a PAN coordinator or their coordinator perform an orphan scan through its list of available channels. Once the coordinator receives the inquiry from the orphaned device, it sends back a coordinator realignment command.

After a channel scan, if there are no appropriate networks to join, an FFD may begin operating as a PAN coordinator (sending beacons) through the use of the MLME-START.request primitive. In some instances, it is possible that two PAN coordinators with the same PAN identifier are collocated. One reason for this problem is that a PAN coordinator may self-assign its PAN identifier without consulting with neighboring coordinators.

A PAN coordinator detects a PAN identifier conflict if it receives a beacon with its own PAN identifier from a coordinator acting as a PAN coordinator or when a *PAN ID conflict notification* command is received from one of its child devices. Similarly, a network device can detect a conflict when it receives a beacon with the same PAN identifier of the network it belongs to, in a time that has no correlation with the beacon period or when this beacon carries a source device address different from the address of its original PAN coordinator. In either case, the network device will send a *PAN ID conflict notification* command to its PAN coordinator.

After the PAN coordinator detects the conflict, it performs an active channel scan that enables the selection of a new PAN identifier and then it proceeds to broadcast a *coordinator realignment* command to its associated network devices.

 The procedure for a PAN coordinator to assign its PAN identifier is out of the scope of IEEE Std 802.15.4.

Device Synchronization

In a beacon-enabled network, devices are required to synchronize with the beacon frames in order to detect any pending messages. In a non-beacon-enabled network, devices may also send beacons acting as "hello" messages in order to enable discovery by neighboring devices.

GTS Management

The PAN coordinator is in charge of maintaining the complete administration of the superframe structure and the control of the allocation, deallocation, and reallocation of guaranteed time slots.

Network devices can request GTSs that extend over several superframe time slots. The PAN coordinator will allow the request depending on the availability of time slots in the contention-free period and the needs of the overall network. The MCPS-DATA.request primitive contains a parameter that indicates if the frame will be transmitted using a GTS or if it will be transmitted using the contention access period. When a GTS transmission is requested, the transmission is deferred until the start of the assigned GTS.

A network device or the PAN coordinator can initiate GTS deallocation. A network device uses the *GTS request* command, while the PAN coordinator indicates any changes to the GTS allocation in its beacon. A PAN coordinator initiates GTS reallocation after detecting GTS fragmentation due to previous GTS deallocations.

SECURITY SERVICES

The MAC of IEEE Std 802.15.4 provides security services that are controlled by the MAC PIB. The MAC sublayer offers two security modes: an unsecured mode and a secured mode.

To meet the objectives associated with the security modes, a critical function of IEEE Std 802.15.4 MAC is frame security. Frame security is actually a set of optional services that may be provided by the MAC to the upper layers. The standard strikes a balance between the need for these services in many applications, and the desire to minimize the burden of their implementation on those applications that do not need them. The available services are described in the following sections.

 Due to the variety of applications targeted by IEEE Std 802.15.4, the processes of authentication and key exchange are not defined in the standard.

Data Confidentiality

Data confidentiality is provided through encryption using a symmetric cipher, in which the same key is used to encrypt plaintext at the message source and then decrypt the resulting ciphertext at the message destination. Devices without the key cannot decrypt the message. The standard defines encryption of beacon payloads, command payloads, and data payloads; other messages, such as acknowledgments, and message components, such as addresses, are not encrypted.

Data Authenticity

Data authenticity, also called *data integrity*, is a service that enables a receiving device to detect the modification of a message by parties without the correct cryptographic key, by appending a message integrity code (MIC) to the message. The standard defines integrity checking of MAC header, auxiliary security header, and unsecured payload fields of data, beacon, and MAC command frames. Integrity is expected to be the security service most used by IEEE Std 802.15.4 applications. Unlike file transfer or voice communication applications common with other protocols, IEEE Std 802.15.4 applications often transmit messages that do not convey secret information. What is often more important is that the messages are authenticated (i.e., the source is known and trusted), and that any tampering with or modification of the message is detected. For example, the fact that a light in a building is turned off is usually not secret; however, it is more important to be sure that any message to the light instructing it to turn off comes from a trusted source (e.g., the room switch), rather than a prankster or cracker.

To avoid replay attacks to an IEEE Std 802.15.4 network, in which an old message is stored by a malicious entity without the cryptographic key and then replayed later, the data authentication service places one of an ordered sequence of values in the auxiliary security header. When received, this value is compared against a stored value; if it is newer than the stored value the replay protection check passes and the new value is stored. Although the replay protection can determine that one message is newer than another, it is a relative determination; no statement of absolute time is made. The use of the data authentication service requires additional memory to store the present value, and increased latency to transmit the message integrity code.

IEEE Std 802.15.4 provides combinations of these three services in two security modes, which are designed to serve a wide range of applications.

> The security service of Std IEEE 802.15.4 has been simplified and streamlined in the 2006 revision. The result is that the secured mode is not backward compatible, however communication between devices following the 2003 and 2006 editions is still possible by using the unsecured mode.
>
> **IEEE 802.15.4 2006**

Unsecured Mode

In unsecured mode, no security services are provided. This mode is suitable for some applications in which implementation cost is important, and security is either not required or obtained in other ways. Examples of this type of application include an advertising kiosk in a public place, or a low-power wireless motor controller in the center of a large, physically secure area outside of which its signal cannot be detected.

Secured Mode

> When a secured mode with data authenticity is used, replay protection is always provided.
>
> **IEEE 802.15.4 2006**

In secured mode, the device may offer two security services (data confidentiality and data authenticity), depending on the security level employed. Eight security levels have been defined as shown in Table 5–2. The unsecured mode is defined as level 0. This level provides the backward compatibility with deployed devices following the 2003 edition of the standard.

The AES-128 (Advanced Encryption Standard, with 128-bit keys and 128-bit block size) symmetric-key cryptography algorithm [24] is employed for security levels one to seven.

Security levels one to three provide the data authentication service with integrity codes of 32, 64, or 128 bits in length to provide data authenticity and replay protection. Security level four provides data confidentiality service, while security levels five to seven employ AES to provide data confidentiality and data authenticity (with message integrity codes of either 32, 64, or 128 bits). The length of the

Table 5–2: IEEE Std 802.15.4 security levels

Security Level	Security Attributes	Data Confidentiality	Data Authenticity	Replay Protection
0	None	OFF	NO	NO
1	MIC-32	OFF	MIC-32	YES
2	MIC-64	OFF	MIC-64	YES
3	MIC-128	OFF	MIC-128	YES
4	ENC	ON	NO	NO
5	ENC-MIC-32	ON	MIC-32	YES
6	ENC-MIC-64	ON	MIC-64	YES
7	ENC-MIC-128	ON	MIC-128	YES

integrity code does not refer to the strength of the AES algorithm, but rather to the number of bits of the code actually transmitted with the message.

This set of security suites offers several advantages. A major advantage is that it employs only one cryptographic algorithm for all levels of security. Other alternatives require additional algorithms, (for example, a hash algorithm for integrity checking), to provide a complete multilevel security solution. This results in a larger and, therefore, more costly implementation. In particular, by reusing AES in a clever way, the AES-CCM* modes enable a single algorithm to provide both security services in a very small implementation. The CCM* is an extension of the counter with cipher block chaining message authentication code mode of operation. The use of AES-CCM* modes retains compatibility with other IEEE 802 standards, such as IEEE Draft Std 802.11i and IEEE Std 802.15.3. This not only allows circuit and code reuse between the standards; it ensures that a larger audience is reviewing and studying the security algorithms for weaknesses. Over time, this high level of interest and analysis increases confidence in the security algorithms.

Chapter 6 Network Functionality

...getting information from source to destination

IEEE Std 802.15.4 was designed to support multiple network topologies—from a star network to many network types based on peer-to-peer communication, including mesh, tree, cluster, and cluster-tree networks. Although the network layer *per se* is beyond the scope of the standard, this chapter describes some sample networks that may be constructed using IEEE Std 802.15.4, including samples of their associated message routing algorithms. Although message routing algorithms are beyond the scope of the standard (except for the simple MAC switching functionality of the star network), these algorithms are described in this chapter to illustrate the type of performance that may be achieved by IEEE Std 802.15.4 networks, and the wide variety of applications to which they may be applied.

FEATURE OVERVIEW

IEEE Std 802.15.4 supports star networks in a master–slave configuration and a very general form of peer-to-peer communication, which may be used to construct a variety of network types.

The star network (Figure 6–1) is a single- or double-hop network (i.e., all network devices are in range of a single common device), whereas peer-to-peer networks, for example, mesh (Figure 6–2) or cluster (Figure 6–3) networks, may be multi-hop networks (i.e., the message source and destination devices are not necessarily within range of each other, and communication may occur via multiple intermediate devices that relay messages). The standard thus supports both single-hop and multihop communication links.

Starting a Network

As described in Chapter 5, an IEEE Std 802.15.4 network is started by an FFD when it becomes the PAN coordinator. All networks must have exactly one PAN coordinator in each network. Figure 6–4 shows the network formation, which starts when an upper layer of the FFD sends the MLME-SCAN.request primitive to the MLME, requesting an active channel scan. Upon completion of the scan, the results are sent back to the upper layers via the MLME-SCAN.confirm primitive. If the results of the scan are acceptable, the upper layer selects a PAN identifier (which may be predetermined), and sends the MLME-START.request primitive (with the PAN coordinator parameter set to TRUE) to the MLME.

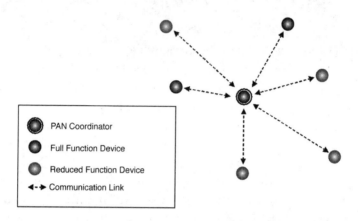

Figure 6–1: Star network

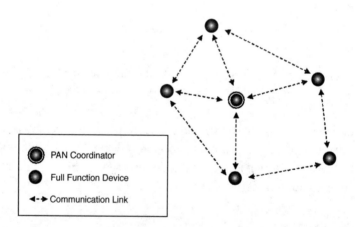

Figure 6–2: Mesh network

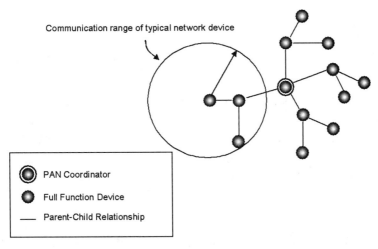

Figure 6–3:   Cluster network

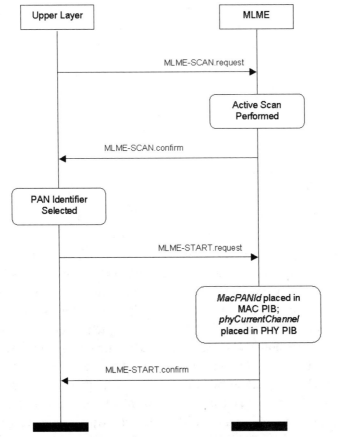

Figure 6–4:   Network formation

The MLME-START.request primitive requires the MAC sublayer to place the PAN identifier *macPANId* value in the MAC PIB and the logical channel *phyCurrentChannel* value in the PHY PIB. After this is completed, the MAC sublayer sends the MLME-START.confirm primitive to the upper layers, the FFD begins operating as a PAN coordinator, and the network is started.

All IEEE Std 802.15.4 devices have a unique 64-bit IEEE address, *aExtendedAddress* (often called a "MAC" or "physical" address), stored in the MAC PIB table. When a device joins a network, it identifies itself by its *aExtendedAddress* value; at the discretion of the PAN coordinator, it may exchange that address for a shorter, network-specific logical address supplied by the PAN coordinator during the network association process.

UPPER LAYER NETWORK FORMATION POLICIES AND ALGORITHMS

As previously indicated in Chapter 5, some policies and algorithms associated with network formation and association are performed by upper layers of the protocol and not defined in the standard. A few alternatives for these algorithms are discussed here. Recall that what follows are only possibilities—other network-forming policies and algorithms could be used as well.

PAN Coordinator Selection

The first step in network formation is the selection of the PAN coordinator. The policy deciding the transmission of the MLME-START.request primitive (with the PAN coordinator parameter set to TRUE) from an upper layer to the MAC is, of course, determined by that upper layer. One may envision several application-dependent scenarios as follows:

- *Dedicated PAN coordinator:* There are some applications, for example, some home security systems, for which it is clear that only one device, the gateway device, needs to be connected to an outside network. This device should be the PAN coordinator. These applications may require the consumer to purchase the entire system as a whole at one time, so that the manufacturer may have complete control over the final network design and behavior. In other situations, the consumer may have the flexibility to purchase network devices individually. This policy may be reinforced by employing only one FFD in the network (as the PAN coordinator) and populating the rest of the network with RFDs. In such a network, exactly one device is eligible to become the PAN coordinator. This must be clearly understood by the consumer user, who may otherwise attempt to make a network with zero, two, or more devices each attempting to become PAN coordinator, leading to unsatisfactory results.

- *Event-determined PAN coordinator:* In other applications, it is desirable to employ a large number of identical devices, any one of which may become the PAN coordinator given an external stimulus (for example, a button press by a user). In this case, every device must be an FFD.

- *Self-determined PAN coordinator:* For these applications, the goal is to establish a network, but it is less important which device in the network is the PAN coordinator. An example of this type of application is a location-determining network. The purpose of this network is to determine the relative location of each network device, possibly by means of a distributed algorithm. In this type of application, there may be no external gateway, so any network device may serve as the PAN coordinator. One way of forming this type of network is to employ a form of power-on-network formation, in which the upper layer of each device instructs the MAC to begin an active channel scan (the needed step prior to network formation) upon device power-up. The first device to complete this scan (which should produce a negative result if the network is not yet formed) and send the result to the upper layer may be rewarded with the MLME-START.request primitive, and become the PAN coordinator. As always, all prospective PAN coordinators must be FFDs.

Selection of the PAN Identifier

Once the channel scan has been completed and the results returned to the higher layers, the results should be evaluated and action taken by the higher layers.

In some applications, it may seem desirable from a security standpoint to fix the desired PAN identifier *a priori*, so that new network devices will be limited to joining only one particular network (that must be already established). Although possible with IEEE Std 802.15.4 networks, this policy should be discouraged, due to the relatively small number of PAN identifiers available and the unpredictable behavior (from the user's perspective) that could result if two networks with identical PAN identifiers were co-located. (There is a procedure for PAN identifier conflict resolution in IEEE Std 802.15.4.) An alternative method is to require all network devices upon association to check the 64-bit extended address of the PAN coordinator, and compare it with the desired address. In this way, association with the correct network, regardless of PAN identifier, can still be guaranteed.

If a new PAN is to be created in the same region as a pre-existing PAN (a case in which a pre-existing or newly placed FFD becomes a dedicated PAN coordinator for the new PAN), the new PAN coordinator must posses a different PAN identifier than all PAN coordinators it is capable of hearing. However, if the goal is for an FFD to become a member of any network, and assume PAN coordinator status only if no PAN identifiers were received during the channel scan, a number of scenarios are possible.

If exactly one desired PAN identifier has been received during the channel scan, the device may choose to attempt to associate with the device transmitting that PAN identifier. If more than one device was heard transmitting a desired PAN identifier, some type of selection algorithm must be employed to determine to which device association should be attempted. This algorithm may consider other information transmitted in the received beacons, previous history (perhaps previous association attempts with a particular device have failed), or may employ a simple deterministic or stochastic (e.g., round-robin, first-in, or random selection) process.

The final scenario, of course, is the case in which no other PAN identifiers were heard and, for application reasons, the device is not permitted to become a PAN coordinator. In this scenario, the device may return to sleep for an application-dependent period, waking later to repeat the active channel scan.

The Use of Beacons

Beacon use is required in IEEE Std 802.15.4 networks for network discovery (i.e., when the network wishes to attract new network devices). Beacons also are useful for coordination and synchronization of an active network (e.g., for GTS control), and in applications for which it is desirable to minimize coordinator-to-network device message latency, because the network device can readily check the message pending field of a beacon to determine when messages are waiting for it.

However, beacons can be disadvantageous in some applications. For example, if no messages are expected from the coordinator to the network device, and only light traffic is expected from the network device to the coordinator (for example, in a wireless light switch), beacon transmission will waste the power of the coordinator, and reception of the beacons will waste the power of the network device. A better solution is to stop the beacon transmissions and have the network device asynchronously transmit its (rare) data frames as they are generated. This approach is especially desirable if the network is in a star configuration, and the PAN coordinator may be externally powered. In this case, the PAN coordinator may operate constantly in receive mode, while the (presumably battery-powered) network device need only expend significant energy when it transmits a state change to the PAN coordinator. As a result, the battery life of the network device may be greatly extended. This is true even if occasional coordinator-to-network device traffic is expected, requiring the network device to poll the coordinator for data.

THE STAR NETWORK

A star network is a good design choice for applications that need to cover a limited physical area, so that a single device (the master, which will be the PAN coordinator) may be in range of all other network devices (the slaves). IEEE Std 802.15.4 star networks must employ an FFD as the PAN coordinator; however, the other network devices may be either FFDs or RFDs. Because star network devices communicate only with the PAN coordinator, they have potential implementation cost savings over peer-to-peer networks, which must store information for every peer with which they communicate. An IEEE Std 802.15.4 RFD, operating as a network device in a star network, can be a good alternative for simple point-to-point applications that require an extremely low-cost implementation.

Because star networks by their nature are single- and double-hop networks, message latency can be lower than in multihop networks. If latency is a critical performance metric of the desired application, a GTS may be assigned by the PAN coordinator to reserve time for a particular network device in each super-frame, avoiding the delay of contention-based channel access and guaranteeing a fixed amount of bandwidth to the network device. In this way, an IEEE Std 802.15.4 star network can provide maximum message latencies as low as 15.36 ms, suitable for use in PC peripheral applications such as wireless mice and joysticks. In the extreme case, it is possible to extend a single GTS to encompass all time between beacons; this enables a single network device to have the entire channel bandwidth, in excess of 115.2 kb/s for the 2.4 GHz band, for relatively high-bandwidth applications.

Conceptually, message routing in star networks is viewed differently than message routing in peer-to-peer networks. Because the PAN coordinator of a star network can hear all network devices and directly controls access to the shared channel, routing in star networks is viewed as occurring in the MAC layer, as packet switching, rather than in the network layer as part of a peer-to-peer message routing algorithm. With this idea in mind, message routing in star networks has been defined in IEEE Std 802.15.4. In fact, routing at this level can be shown to have some implementation advantages, because simple message relaying need not involve the higher layers of the protocol.

PEER-TO-PEER NETWORKS

The peer-to-peer communication capability of IEEE Std 802.15.4 devices allow the creation of many types of peer-to-peer networks, each with their own advantages and disadvantages. A sampling of network types possible with IEEE Std 802.15.4 FFDs is described in the following sections.

The Flat "Mesh" Network Topology

The simplest type of peer-to-peer ad hoc network to envision is perhaps the flat "mesh" network, a network composed of a number of identical network devices, not all of which may be in range of any one device. Messages may be relayed from source devices to destination devices via a large number of routing algorithms.

Although the network devices in the mesh may be identical, one must have special capabilities to perform the PAN coordinator function, give the network its PAN identifier, and control the association of new devices to the network. However, the PAN coordinator need not take a role in message routing.

 A true mesh network has regularly spaced devices forming a repeating grid pattern of communication links; however, the definition used here, allowing an ad hoc device distribution, has become common.

Routing

A principal problem to be overcome in a flat mesh network design is the problem of addressing. Because the network is logically flat (i.e., there is no hierarchy) and there is no other grouping or organization of network devices, the address of a network device does not provide clues to the route needed to get a message to it.

Nevertheless, many routing algorithms have been devised for such networks, a few of which are listed here as follows:

- *Flooding:* The simplest method to route messages is to send a copy of each message to all network devices. Although this is a correct algorithm (i.e., it will in fact deliver the message to the intended recipient), it is not very efficient, especially for the large networks IEEE Std 802.15.4 can support. Much power can be wasted sending the message to unintended recipients. This method is seldom used in practical networks of any size to transmit data traffic, although it is needed in nearly all networks for at least some control and status message traffic.

- *The Bellman-Ford algorithm:* The Bellman-Ford algorithm requires all network devices to maintain a routing table that contains routing costs (usually the number of hops, although more complex cost accounting, such as available power of network devices in the route, can be used) of the optimum route to all other network devices, plus the address of the first device in that route. Devices maintain their tables by exchanging them with all devices within their range, and then comparing entries by destination. The route of a message is determined by the source device in advance ("source routing"), and placed *in toto* in the MSDU, as part of the message payload. Relaying network devices follow these routing instructions to deliver the message. Although an elegant solution to the routing problem, the Bellman-Ford algorithm suffers from poor dynamic behavior. If a communication link is broken, for example, its convergence behavior to the new set of optimal routes is not good. In addition, because the routing tables must have as many entries as there are network devices in the network, the costs to exchange them become prohibitive once the network becomes large.

- *The GRAd (Gradient Routing for Ad hoc networks) algorithm:* The GRAd algorithm [19] requires all network devices to maintain a cost table that lists the routing cost to each potential destination device (as opposed to the Bellman-Ford algorithm which lists all network devices). However, cost tables are not exchanged among network devices, and messages are not sent via a unicast transmission to particular network devices along a route determined by the message source. Instead, messages are broadcast by the source to all network devices in range, along with the cost value to the destination found in the source's cost table. Any neighbor hearing the broadcast message, and having a routing cost for the destination less than that sent by the source (and therefore presumably closer to the destination), waits for a random period and then rebroadcasts the message, listing its own routing cost. Acknowledgment is passive; the acknowledgment to the transmitting device is detection of the retransmission by a neighboring device. All others ignore the message, except to record the routing cost and destination sent by their neighbor, which they use to update their cost tables. In this way, the message slides down a "cost gradient" to its destination. The GRAd algorithm requires each network device that may be a message destination to flood the network upon network association with a special announcement message, so that it will appear in the cost table of other network devices. Alternatively, specialized route request messages can be broadcast by devices with messages for destination nodes for which they have no cost value stored.

- *Ant algorithms:* Because they also are ad hoc and self-organizing, biological analogies to the routing problem have also been explored. Of particular interest is the communication paradigm of ants. Ants communicate by placing a pheromone trail along the ground. If, for example, a wandering ant finds food, it returns to the nest, placing the "food" pheromone on the ground as it does so. Other ants detect this trail, arrive at the food, and return to the nest, laying down pheromone trails of their own that diffuse together into a wide trail. The pheromone evaporates quickly; an unused trail becomes undetectable to other ants in a short time. In this way, the ants quickly develop a single path to the food that is used by almost all ants.

Because this use of pheromones is a distributed communication system in which simple behavior of individual ants results in complex behavior of the larger network, it has clear appeal to the network routing problem, especially in the case where there is a single message sink, but possibly many message sources. One approach is to model the information sink (i.e., the PAN coordinator, the destination of all messages) as the nest, the network devices sourcing messages as the food, the collection of intermediate (relaying) network devices as the ground between food and nest, and the messages themselves as the ants. The PAN coordinator periodically transmits a special announcement message (containing a "number of hops taken" field), which is passed from network device to network device by unicast transmissions. At each network device, the device stores the source identity, and the identity of the neighbor from which it received the message, in a table, along with a time stamp. The network device then increments the "number of hops taken" field in the message and forwards it to a different neighbor. Network devices perform the pheromone "evaporation" function by removing old entries from their tables.

When a "real" message is generated in the network, destined for the PAN coordinator, if the source has not received the special announcement message from the PAN coordinator recently (so that an entry is not present for it in its table), it sends the message to a random neighbor. This process is repeated until the message is sent to a network device that has a table entry for the destination. That device then sends the message to the neighbor from which it received the special announcement message, as recorded in its table. The message is now "on the pheromone trail," and it will eventually arrive at its destination.

The ant communication paradigm has appeal for particular networks in which it can be guaranteed that there will be only a few network devices acting as message destinations, because each destination must transmit regular announcement messages. Examples of such networks include wireless sensor networks, which typically have many source devices (sensors) and few destination devices.

The Cluster Network Topology

The previous algorithms notwithstanding, routing in a flat network has unavoidable limitations for many applications, and it would be desirable to have a network in which the routing algorithm was assisted in some way by the logical structure of the network, especially as the number of devices in the network becomes large.

(Note that the *physical* structure of the network may still be ad hoc, however). This could eliminate, for example, much of the announcement message passing of the ant algorithm, and the exchange of very large routing tables inherent in the Bellman-Ford algorithm.

One approach to this problem suitable for many applications is the cluster network. In a cluster network, there is the concept of a "parent–child" relationship between network devices (Figure 6–5). The network forms, as do all IEEE Std 802.15.4 networks, with the PAN coordinator as the first device in the network. When a new device associates to the PAN coordinator (and therefore to the network), it becomes the child of the PAN coordinator, and the PAN coordinator becomes the parent of the new device. Should a second device come into range of the first network device (but, perhaps, out of range of the PAN coordinator), the second device may join the network as a child of the first device. Network devices may have many children (and grandchildren), but only one parent (Figure 6–6).

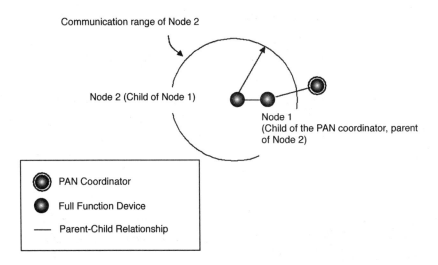

Figure 6–5:   Formation of a cluster network

Discovery of the network by new devices is made possible by the transmission of beacons by each network device (including the PAN coordinator). When a new device appears, it may hear several beacons, and have several potential parents from which to choose. If (as is often the case) it is desirable to connect network devices as close to the PAN coordinator as possible, information may be placed in the beacon payload to assist the prospective network member in the parental selection process.

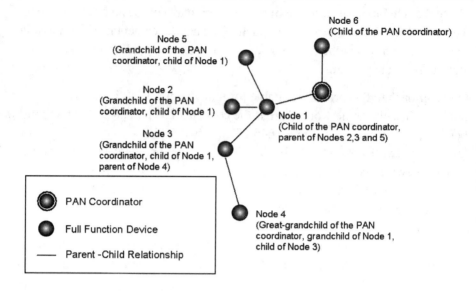

Figure 6–6: Parents, children, and grandchildren in a cluster network

The structure of the cluster network is controlled to some extent by the PAN coordinator, which retains authority over network association (regardless of which network device a prospective member may contact). The PAN coordinator may prohibit, for example, a prospective member from joining the network at a device distant (many hops away) from the PAN coordinator, while allowing other prospective members to join closer to the PAN coordinator, in order to encourage a "flatter" cluster structure and control the latency of messages sent in the network.

One advantage of cluster networks is that it is straightforward to perform a periodic network status update, so that the PAN coordinator can be made aware of any broken communication links or missing network devices anywhere in the network. One way this can be accomplished is to have the PAN coordinator generate a "status update request" message, sent to all network devices that are not parents (i.e., all devices at the ends of the cluster branches). When received, every parent device relays this message to all of its children, so that all network devices either receive or relay this message. When a network device without children receives the status update request message, it replies with a "status update response" message, a very short message containing its network address sent to the PAN coordinator. Intermediate devices on the return path to the PAN coordinator "eavesdrop" on this message as they relay it, gaining information about network devices in their cluster, but perhaps out of range. They then append their information to the message (perhaps consolidating the messages received from several children) and forward it to their parent. This procedure is shown in Figure 6–7.

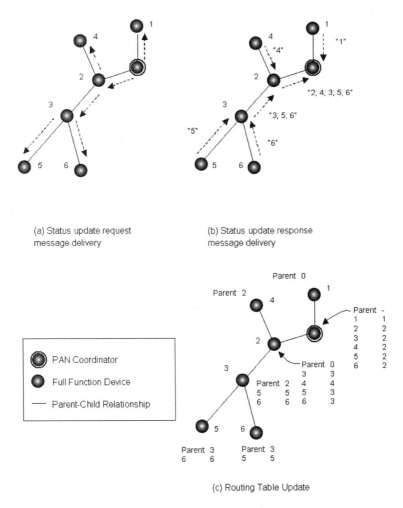

(a) Status update request message delivery

(b) Status update response message delivery

(c) Routing Table Update

Figure 6–7: Network status update messages and routing in a network

A variant of this procedure is to require the status update response message to include the complete routing table of each network device. While lengthening the response message, this can make the PAN coordinator (and, in fact, any eavesdropping network device) aware of alternative routes to the same destination device, which may improve network reliability in the event of a network device or communication link failure.

Routing

Because their network devices have an understanding of network connections in their local area, routing in cluster networks can be more efficient than in flat networks. One efficient way to route in cluster networks is to use information gained in eavesdropping the status update response messages. By doing so, eavesdropping devices have a list of network devices "downstream" of the device from which the status response message was received. This information can be stored in a routing table.

A general routing algorithm for cluster networks is shown in Figure 6–8. The algorithm assumes that the network device stores a routing table, which consists of entries from two sources:

- Entries added by eavesdropping on devices within reception range. One of these entries will be for the parent of the device. These entries make up the "neighbor list," the list of all devices to which messages may be passed without intermediate relay.

- Entries added by eavesdropping on network status reply messages. These messages inform the device of all devices in the network "downstream," that is, away from the PAN coordinator.

Following the algorithm shown in Figure 6–8, a network device checks first if the destination device has an entry in its routing table. If an entry is found, the message is passed to the appropriate device. If an entry is not found, the message is sent to the parent, which, because it is closer to the PAN coordinator may have additional routing information by virtue of eavesdropping on network status response messages from a larger number of devices. In the extreme, a message may be relayed all the way to the PAN coordinator, but delivery is assured nevertheless.

A significant advantage of cluster networks for wireless sensor network applications is the small size of the network device routing tables, compared to those of flat networks. In general, for flat networks, a table entry is required for every potential destination; for a cluster network, the routing table may be much smaller because destinations not found in the table are still routed, via the parent. A small table size reduces the memory requirement of the network devices, which can be a major factor in their product cost.

A disadvantage of cluster networks is the non-uniform distribution of message traffic among the network devices. Some devices, especially those logically close to the PAN coordinator, may have significantly higher traffic than devices further away from the PAN coordinator, leading to unequal battery life among network devices. Should the power supply of a device become exhausted, a network partition could occur. Several methods have been proposed to mitigate this effect, including the rotation of PAN coordinator duties among alternative network devices and the use of multiple routing table entries for the same destination.

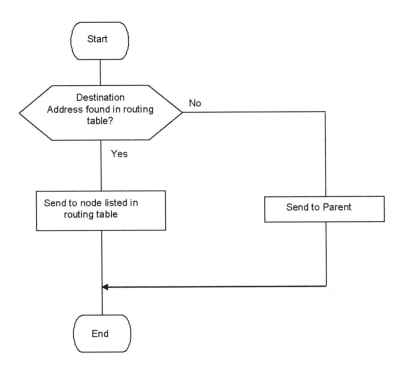

Figure 6–8:   A message routing algorithm for cluster networks

The Cluster Tree Network Topology

The address space of IEEE Std 802.15.4 MAC is capable of supporting a large number of network devices. However, as the number of network devices continues to grow, even the routing tables of devices in a cluster network may grow to an impractical size, because a device's routing table must contain an entry for each of its offspring (child, grandchild, etc.).

To address this issue, hierarchy may be employed. A large network may be broken up into several, smaller clusters, connected in a hierarchical tree. An example of such a "cluster-tree" network is shown in Figure 6–9.

The large network shown in Figure 6–9 is composed of four smaller clusters, each with a "cluster head," or coordinator. The coordinator for Cluster 0 is, of course, the PAN coordinator. If the PAN coordinator has a set of rules governing network formation, the clusters may be formed as a natural part of the network formation process. These rules may take the form of a limit on the number of devices in a cluster, a limit on the number of hops any device may be from a coordinator, or a more sophisticated algorithm. Should the attempted association of a candidate device to a network device violate these rules, the PAN coordinator may allow the device to join, but only as the coordinator of a new cluster.

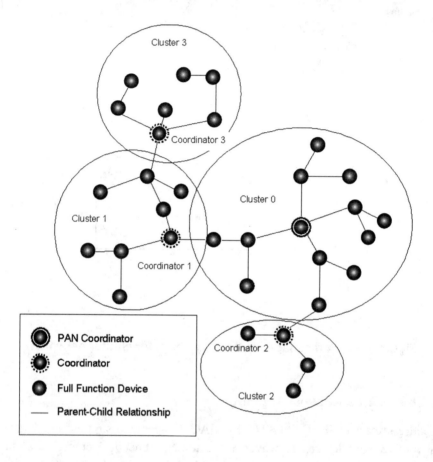

Figure 6–9: The cluster-tree network

The logical, short address of a network device in a cluster-tree network is now a hierarchical address consisting of two parts, a cluster identifier and a network device identifier. The network device identifier of a coordinator is always zero; the cluster identifier of the PAN coordinator is always zero.

In the cluster-tree network, the PAN coordinator may still transmit network status request messages. One way to do this is to have the network status request messages route, as in the cluster network, to the ends of the branches (the "leaves" of the tree). The network status response message is also handled in the same way, except that the device identifiers in the message are deleted by the eavesdropping coordinators; i.e., network status response messages leaving the coordinators contain only the identity of the cluster. This takes advantage of the address hierarchy to reduce the size of the network status response message, which, in a large network, may otherwise be impractical.

Note that, as in the cluster network, information extracted from the routing table of each network device may be included in the network status response messages to exploit routing redundancies.

Routing

A possible routing algorithm for a cluster-tree network is shown in Figure 6–10. In this algorithm, a device first checks to see if the cluster identifier of the destination is in its routing table. If not, the destination is an unknown cluster and the device routes the message to its parent.

However, if there is a cluster identifier match in the routing table, the device then checks those table entries with a cluster identifier match for a network identifier match. If one is found, the message is routed according to the table entry. If one is not found, the destination is an unknown network device in a known cluster, and the device must choose which network device to use to relay the message to the cluster by some set of arbitration rules.

If the cluster identifier of the destination is the same as that of the network device attempting to route the message, as in the cluster network, the message should be routed to the parent of the network device. However, the arbitration rules governing the relay of messages to other clusters can be significantly more flexible, and may consider, for example, the time a routing table entry was made—routing the message via the network device most recently added to the table, in the belief that the route is the most up-to-date and most accurately reflects the present state of the network. Other considerations are possible, many of which are optimal for particular applications.

THE NETWORK TOPOLOGY DECISION

Because IEEE Std 802.15.4 is capable of supporting a large number of ad hoc network topologies, an application designer employing IEEE Std 802.15.4 must select an appropriate network topology for the intended application. Further, if the selected network topology is a peer-to-peer topology in which multihop communication is planned, the routing algorithm must also be selected.

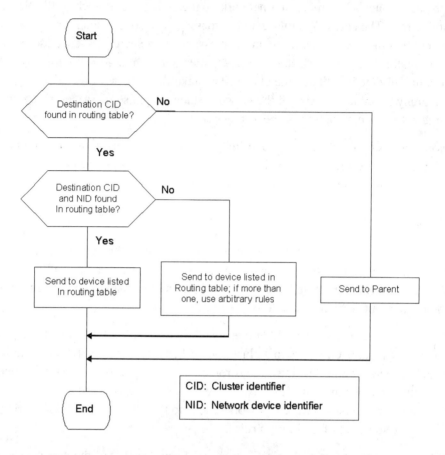

Figure 6–10: A cluster-tree routing algorithm

As an aid to this decision-making process, Table 6–1 summarizes the major network types discussed in this chapter, including their strengths, weaknesses, and potential applications. Note that Table 6–1 is only a guide, intending to establish general trends; almost any of these networks may be used successfully in almost any application, albeit with variations in network performance parameters such as message throughput and latency, and network device performance parameters such as power consumption.

 The network types listed in this section are only examples; many other topologies and routing algorithms are possible.

Table 6–1:   Network topology comparison

Network Type		Strengths	Weaknesses	Possible Applications
Star		Low message latency; centralized network control	Can cover only a limited physical area (single-hop communication)	Home automation; PC peripherals
Peer-to-Peer		Can cover a large physical area (multi-hop communication)	Higher message latency	Wireless sensor networks; industrial control and monitoring
	Flat	Simple network devices	Does not scale well as the number of potential destination devices increases	Wireless sensor networks
	Cluster	Supports a larger number of potential destination devices	Uneven power consumption among network devices	HVAC systems
	Cluster Tree	Can support very large networks	Network maintenance overhead	Industrial control and monitoring

Part III

Chapter 7   System Design Considerations

...a systemic view

IEEE Std 802.15.4 was created to address applications that could not be satisfied with existing WLAN and WPAN communication protocols, primarily due to the cost, size, and/or current drain of their implementations. This chapter discusses system design considerations needed to take advantage of the capabilities of IEEE Std 802.15.4 and achieve low cost, size, and current drain implementations of the standard.

To reduce cost, size, and current drain of its implementations, IEEE Std 802.15.4 was designed for a high level of integration, with a minimum number of external parts. Several features in the protocol help achieve this goal.

Direct Sequence Spread Spectrum

The use of Direct Sequence Spread Spectrum (DSSS) enables an IEEE Std 802.15.4 implementation to be largely a digital one, with only a relatively small number of analog circuits. This feature "future-proofs" the standard; the evolution of the standard as improvements in integrated circuit lithography will increase digital circuit density, leading to even lower implementation costs. DSSS offers other implementation advantages. The DSSS processing gain provides rejection to interfering signals, reducing the requirements on the channel filter. The ability to easily design orthogonal, multilevel signals with DSSS enables the IEEE Std 802.15.4 2.4 GHz physical layer to have simultaneously a relatively fast data rate of 250 kb/s (to finish transmission and return to sleep quickly) and a relatively low symbol rate of 62.5 kSymbols/s (to minimize current drain while active); both of these features improve average power consumption. Further, in some regulatory jurisdictions, some type of spreading is required, and DSSS enables faster network discovery and synchronization than does Frequency Hopping Spread Spectrum (FHSS). For these reasons, DSSS was chosen over FHSS and narrowband techniques.

High Channel Separation/Modulation Bandwidth Ratio

IEEE Std 802.15.4 has a high ratio of channel separation to modulation bandwidth. That is to say, adjacent channels are far away from the desired signal, compared to the width of the desired signal (see Figure 7–1). The first null in the

---

transmitted IEEE Std 802.15.4 spectrum in the 915 MHz band is 600 kHz away from the signal center, with 1.4 MHz separation to the edge of the adjacent channel signal; in the 2.4 GHz band the first null is 1.5 MHz away from the signal center, with 3.5 MHz separation. In both cases, the separation-to-bandwidth ratio is $1.4/0.6 = 3.5/1.5 = 2.33$. However, IEEE Std 802.15.4 requires very little selectivity to the adjacent channel, only the alternate channel, resulting in an effective ratio of $3.4/0.6 = 8.5/1.5 = 5.67$. In comparison, the transmitted IEEE Std 802.11b spectrum, in which the first null is 11 MHz away, with 14 MHz to the edge of the next nonoverlapping adjacent channel, has a ratio of only $14/11 = 1.27$; IEEE Std 802.11b requires 35 dB of selectivity to this channel. IEEE Std 802.15.4 channel filter stopbands are therefore well below the desired passband.

This has several beneficial effects:

- Receiver types that employ integrated selectivity, such as low-IF and zero-IF architectures, are possible. Discrete surface acoustic wave (SAW), ceramic, or crystal filters are not needed. This reduces implementation size and cost.

- For low-IF and zero-IF receiver implementations, the low-pass cut-off frequency of the channel filter may be raised. This reduces the size and cost of the integrated filter.

- Fewer poles of channel filtering may be used. This also reduces the size and cost of the filter.

- With fewer poles of filtering, and a higher corner frequency, the time required to settle turn-on transients in the channel filter may be reduced. This enables a faster receiver warm-up, minimizing battery power lost during the warm-up period.

- The relatively large channel separation means that a high synthesizer reference frequency may be used. Because synthesizer lock time is related to the frequency at which the synthesizer's phase detector operates, a higher reference frequency enables shorter synthesizer lock times, which allows shorter warm-up times and therefore improves battery life.

Relaxed Transmitted Error Vector Magnitude (EVM) Requirements

The relaxed receive selectivity and transmitted error vector magnitude (EVM) requirements of IEEE Std 802.15.4 allow the sideband noise specifications of the receiver local oscillator and transmitter oscillator to be relaxed, perhaps to the point that a very low-cost, fully-integrated oscillator may be employed. This integrated oscillator, which could be of the inductor-capacitor (LC) or (preferably) the resistor-capacitor (RC) type, would eliminate the large and costly external voltage controlled oscillator (VCO) resonator, and other associated passive components, that would otherwise be needed.

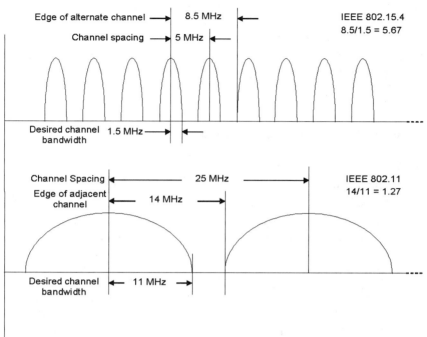

**Figure 7–1: Comparison of IEEE Std 802.11 and
IEEE Std 802.15.4 2.4 GHz PHY selectivity requirements**

Constant-Envelope Modulation

A major factor leading to reduced transmit power consumption is the use of
constant-envelope modulation in the form of half-sine-shaped O-QPSK in the
high band. Transmitters must be designed, of course, to transmit the peak envelope power required by their modulation format; for simple circuits suitable for
IEEE Std 802.15.4 implementations, this requires the amplifier dc bias power
(which can be a significant fraction of the total transmitter power consumption) to
be proportional to the peak envelope power. The *average* envelope power, on the
other hand, is the significant parameter when determining range (given a fixed
data rate). If a modulation format is chosen such that the peak envelope power is
greater than the average envelope power, dc bias power must be spent in the
transmitter commensurate with the higher peak power, even though the range
achieved is that of the lower average power—clearly an undesirable situation for
power-sensitive applications. (Such modulation formats, however, may have
advantages, such as spectral efficiency, that make them suitable for other
applications.) IEEE Std 802.15.4 therefore employs a modulation format having a
constant envelope, in which the peak and average envelope powers are equal (a
peak-to-average power ratio of one). An implementation of IEEE Std 802.15.4
may take advantage of this to employ a simple and relatively efficient transmitter.

Relaxed Receiver Maximum Input Level

The IEEE Std 802.15.4 receiver maximum-desired signal input level specification, –20 dBm, is significantly less than that of other WLAN and WPAN standards, and it enables an IEEE Std 802.15.4 system designer to reduce the current drain of the receiver front end accordingly.

TIME AND FREQUENCY REFERENCE TRADEOFFS

The design of the time and frequency reference subsystem of an IEEE Std 802.15.4 implementation can have a marked effect on overall system cost, size, and current drain. One possible approach to the design of a stand-alone IEEE Std 802.15.4 implementation employs two reference crystal oscillators, a high-frequency oscillator for the RF frequency reference and a low-frequency oscillator for the protocol time base. In this design, the low-frequency (and therefore lower power) crystal oscillator for the time base runs constantly, providing protocol timing information. The high-frequency (and therefore higher-power) crystal oscillator for the RF frequency reference is then started immediately prior to a transmit or receive period and is turned off immediately afterward, thereby minimizing its power consumption. If the required warm-up time of the RF frequency reference is low enough so that it is not a significant factor in total energy consumption for the application under consideration, this approach can produce low average power consumption, at the cost of requiring two external crystals.

A second approach employs only a single, high-frequency crystal oscillator. In this scenario, the oscillator is used for both the RF frequency reference and the protocol time base, and it is therefore never turned off. (The oscillator output is divided down to a low frequency to be used as the protocol time base.) When needed as the RF frequency reference, its output is used directly; because it is already operating, no warm-up period is needed and therefore no power is lost there. This second approach is lower cost than the first, because only one crystal is needed; however, because a higher-frequency oscillator, plus frequency division circuitry, is used for the time base, the average power consumption may be higher. This trade must be evaluated by the system designer based on the expected applications of IEEE Std 802.15.4 implementation.

SINGLE-CHIP VERSUS MULTIPLE-CHIP IMPLEMENTATION

An integrated circuit (component) manufacturer entering the IEEE Std 802.15.4 market is faced with the decision of whether its design should be a two-chip design, in which IEEE Std 802.15.4 RF transceiver is on a separate chip from the protocol handler (Figure 7–2*a*), or a one-chip design, in which the RF transceiver and protocol handler are integrated on the same die (Figure 7–2*b*). A one-chip design, because it has the highest level of integration and fewest number of IC packages, may have the smallest implementation and lowest component cost, and it may be optimum for stand-alone applications that do not have an available host processor (as described in the following sections). However, a two-chip design may employ more optimal IC processes that are better suited to the circuit requirements (e.g., BiC-MOS for the RF transceiver and CMOS for the protocol handler), and it could result in a reduced time to market compared to a chip in a compromise integrated circuit (IC) process chosen for the one-chip design. In addition, if a wide product line of stand-alone IEEE Std 802.15.4-enabled products is envisioned, a two-chip design may allow the substitution of multiple processor types as the protocol handler IC, having different amounts of memory and capable of economically supporting a range of applications of varying complexity. The two-chip design also allows flexibility in the choice of RF transceiver with a single protocol handler; for example, one may want to enable either a dedicated IEEE Std 802.15.4 RF transceiver or one capable of servicing multiple protocols, for example both IEEE Std 802.15.4 and IEEE Std 802.11a. A compromise alternative to this decision is to employ two IC die, but placed in a single IC package.

OEM IMPLEMENTATIONS

The system design decisions faced by an original equipment manufacturer (OEM) are equally interesting. One scenario involves the addition of IEEE Std 802.15.4 capability to a larger system; for example, when a general-purpose host processor is available to run an IEEE Std 802.15.4 protocol stack, and a dedicated IEEE Std 802.15.4 protocol-handling processor is not used. In this case, shown in Figure 7–2*c*, the cost of the dedicated processor may be saved by employing a smaller chip performing only IEEE Std 802.15.4 RF transceiver functions—half of the "two-chip" design just described. However, software integration is made more difficult, especially if the host processor is running other real-time applications, because the performance of many protocol functions cannot be delayed. Interrupt handling and context switching, for example, must be carefully evaluated to ensure that enough processing capability [e.g., million instructions per second (MIPS), random access memory (RAM)] is always available, and the throughput and latency performance of the communication link between the IEEE Std 802.15.4 transceiver and host processor must be understood. Often the worst-case scenario, when the host processor is most heavily loaded, is difficult to predict and test, especially for embedded systems. If the software integration problem becomes untenable, a one-chip IEEE Std 802.15.4 implementation may be used to off-load the host processor, as shown in Figure 7–2d.

Lightly loaded host processors, on the other hand, may have little difficulty meeting the real-time timing requirements of the protocol stack. However, unless specifically designed for the task, the host processor may not perform the bursty computation requirements of the IEEE Std 802.15.4 protocol in a power-efficient manner; the average power consumption of the resulting IEEE Std 802.15.4 implementation may suffer as a result. The system designer may achieve the best compromise between implementation cost and power consumption by judiciously placing only those stack components on the host processor that best fit the needs of the intended application.

IEEE Std 802.15.4 was designed to be compatible with low-cost 8051- or HC08-based microcomputers operating at a bus speed of only a few megahertz. This design involved a trade between the utility and flexibility of multiple protocol layers, and the software size reduction that occurs when the protocol is defined as a single layer. The compromise reached was to keep the layered architecture, to retain compatibility with other IEEE 802 standards, but simplify each protocol layer as much as possible.

TIME AND POWER MANAGEMENT

IEEE Std 802.15.4 is designed to support a time base with a tolerance as great as ±40 ppm. This allows the use of very inexpensive reference crystals, minimizing cost. At 2.4 GHz, the standard also supports a beacon interval of $2^{14} \times 0.01536s = 251.65824s$, or over four minutes. This, combined with a minimum beacon PPDU length of 544 μs, enables a transmit duty cycle of $(544 \times 10^{-6})/251.65824 = 2.16 \times 10^{-6}$. Lower values are, of course, possible if beacons are not used at all.

However, a *receiving* device must be receiving at the time the beacon is sent, a time that is uncertain due to the imperfect accuracy of both transmit and receive time bases (Figure 7–3). Without time base regulation, the receiver must receive for a period of $2\varepsilon T_{beacon} + T_c$ s every T_{beacon} s, where ε is the sum of the transmit time base tolerance ε_{Tx} and receive time base tolerance ε_{Rx}; the lowest possible average duty cycle of a receiving device is therefore

$$2\varepsilon + \frac{T_c}{T_{beacon}}.$$

It is clear that, no matter how long the beacon period T_{beacon} is made, the duty cycle is limited by the attainable time base stability ε. (Note that neither transmit

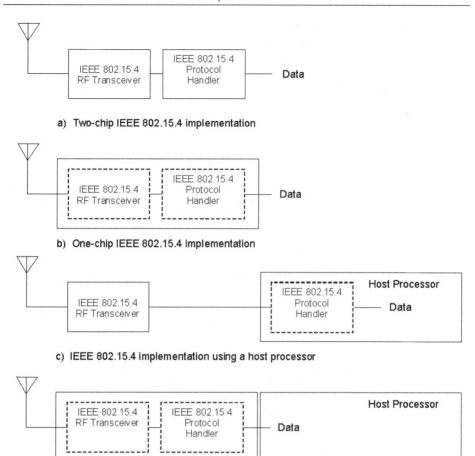

a) Two-chip IEEE 802.15.4 implementation

b) One-chip IEEE 802.15.4 implementation

c) IEEE 802.15.4 implementation using a host processor

d) Alternative IEEE 802.15.4 implementation using a host processor

Figure 7–2: Diverse IEEE Std 802.15.4 hardware implementations

nor receive warmup times are included in this analysis.) For 40 ppm timebases at both transmitter and receiver, this effect becomes significant for $T_{beacon} > 0.5$ s or so; to achieve the very low duty cycles (and resulting long battery life) of which devices employing IEEE Std 802.15.4 are capable, an implementation may therefore require a more stable time base than that specified in the standard. Although the simplest way to do this may be to replace, for example, a 40 ppm crystal oscillator with one of greater stability, this would usually increase the cost of the time base significantly—especially if stabilities below 5 ppm were required. A lower cost approach may be developed by recognizing that the time variation between sequential beacons from the same device is likely to be highly correlated. It is then possible for the receiving device to reduce the device duty cycle by tracking the relative received beacon time from beacon to beacon, and therefore employing a reception window smaller than $2\varepsilon T_{beacon} + T_c$.

Alternatively, a time base drift specification, limiting the allowable time difference between successive beacon periods, could have been specified in IEEE Std 802.15.4, as has been done in other standards. It is clear that a drift specification is not needed for *macBeaconOrder* = 0; when T_{beacon} = 15.36 ms, any possible drift is insignificant when compared to the beacon PPDU length (544 μs at 2.4 GHz) plus warmup time (a total of at least, say, 600 μs). The "drift time," even assuming an 80 ppm difference between devices, does not become significant when compared to 600 μs until one reaches *macBeaconOrder* = 7 (1.96608 s), when it is 157 μs.

It is also clear that a drift specification is not useful for *macBeaconOrder* = 14; a device could move from hot to cold during the four-minute beacon period, and the warm-up time estimate could only be improved by improving the accuracy of the time base, an undesirable option. Very small, single-chip IEEE Std 802.15.4 implementations (perhaps with button cell batteries) may have very low thermal inertia; when moving between temperature extremes during a 16-second beacon period (*macBeaconOrder* = 10), for example, they could well vary by +/–40 ppm, so a drift specification could not be applied here, either, without adding a signifi-cant (i.e., costly) restriction on the hardware. In this regime, *macBeaconOrder* = 10 to 14, the receiving time must be specified based on the time base accuracy specification.

This leaves a small window in which a drift specification would be useful, *macBeaconOrder* = 7 to 9. It was decided that specifying drift only for three *macBeaconOrder* values was not appropriate for a simple protocol with low implementation cost, so a time base drift specification is not included in IEEE Std 802.15.4.

The previous discussion is an example of the importance of time and power management in the design of IEEE Std 802.15.4 implementations, and the con-cern for them that went into the design of the standard. Note that, because the standard is designed to support duty cycles of 0.1% or less, the standby power consumption of the implementation can be the dominant factor in average power consumption (and therefore battery life). It is therefore rewarding to pay attention to standby power consumption during the design process. In addition, because active times are brief (as short as 544 μs for a minimum-length 136-bit beacon PPDU at 2.4 GHz), warm-up times can be a significant fraction of overall active time and can therefore affect the minimum obtainable duty cycle. This effect is important if maximum battery life is to be achieved.

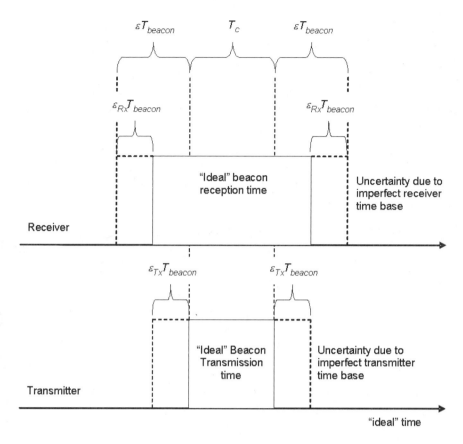

Figure 7–3: Effect of imperfect transmitter and receiver time

If the application can support a star network configuration with a mains-supplied PAN coordinator, and especially if the PAN coordinator is to be the destination of all network traffic (for example, in a lighting network, in which several wireless switches control a lamp PAN coordinator), the other network devices may have an arbitrarily low duty cycle by operating the network without a beacon. In this mode, the PAN coordinator is receiving constantly; the network is in a constant contention access period.

The switches need not transmit or receive at all until some event (e.g., the toggling of a switch) causes a message transmission from a switch to the lamp. Because a lamp is typically connected to the mains, having a near-100% receive duty cycle is not an onerous requirement in this application, and it enables the designer to craft a system in which the switch battery life may be exceptionally long.

ANTENNAS

Because many IEEE Std 802.15.4 products are expected to be physically small, and very sensitive to product cost and power consumption, antennas can be very important to a successful design.

The physically small antenna required in many applications may be inefficient unless careful attention is paid to the design of both the antenna and its placement in the product. An inefficient antenna can significantly reduce both the transmit and receive range, something that can be compensated for only by increasing transmit power and receiver sensitivity—both options that can significantly reduce battery life.

There is a trade to be made among the physical size, instantaneous fractional bandwidth BW, and the maximum achievable radiating efficiency η of electrically small antennas (i.e., those with maximum physical dimensions significantly less than their operating wavelength λ). Specifically [21],

$$\eta = \frac{2\left(\frac{2\pi r}{\lambda}\right)^3}{BW}$$

where $2r$ is the maximum dimension of the electrically small antenna (i.e., the diameter of the smallest sphere that can completely contain the antenna). From this equation, two important relationships may be noted:

- The antenna efficiency is proportional to the ratio of the antenna size and the operating wavelength, *cubed*. This means that, all else equal—including the physical size of the antenna—operation at a higher frequency (a lower wavelength λ) improves antenna efficiency, as does increasing the size of the antenna while keeping the frequency of operation constant. If, however, the antenna size scales with wavelength (as is the case for a half-wave dipole, for example), its efficiency is unaffected by the frequency of operation.

- Attainable fractional bandwidth BW is inversely proportional to attainable efficiency η. In most applications, the efficiency of even well-designed small antennas for any of IEEE Std 802.15.4 bands is low enough that the bandwidth required is easily achieved.

In addition to the antenna size $2r$, it must be recalled that strong circulating (i.e., non-radiating) fields exist near an antenna. These fields vary as d^{-2} and d^{-3}, where d is the distance from the antenna. The fields are therefore very strong close to the antenna, and significant out to a radius of at least $\lambda/10$. Any material in this region, including batteries, circuits, displays, and so on, produces additional loss that can significantly affect antenna efficiency. One therefore should plan on a "keep out" region in the product design of 1.2 cm or so for 2.4 GHz operation, and approximately 3.5 cm for 868 MHz operation; the more material in this region, the lower the antenna efficiency will be. When keep out regions of this size are not possible, the increased loss should be considered in the implementation design. Note that the effect of nearby materials can counteract the increased efficiency of a larger antenna over a smaller one, if the larger antenna must be placed near a lossy object while the smaller antenna is kept away from it.

This theory must be matched with the practical realities of antenna design for small, low-cost products. Because any volume taken by an antenna inside a product is not available for other uses (the placement of circuits, batteries, etc.), it is nearly always desirable to minimize it. Several factors affect this product design decision:

- *The frequency of operation:* As stated previously, for constant efficiency the antenna size scales with the operation wavelength; operation in the 2.4 GHz band would therefore consume less antenna volume.

- *Products for multiple bands:* If a single product design is intended to be used for both 868 MHz and 2.4 GHz operation, for example, space must be made available for the larger 868 MHz antenna.

- *Internal noise sources:* Due to their proximity to the antenna in a small product, even relatively weak noise sources can play havoc with a wireless transceiver. Typical sources are digital circuits (especially the busses of microprocessors and microcomputers, and serial data interfaces) and switching power converters. These components, and the circuit board traces associated with them, should be placed as far as possible from the antenna, something that becomes more difficult as the antenna size increases. When possible, of course, it is desirable to arrange operation of the digital circuits so that they are not clocked when the transceiver is active.

- *Use of external antennas:* In general, external antennas (those outside the product housing) perform better electrically; they are typically away from lossy materials and have fewer size restrictions, so they may be larger. If a directional antenna is needed (perhaps for a long-range application), an external antenna is often the best choice; they may also be required in applications such as wireless home appliances (e.g., a refrigerator) in which the radio transceiver is enclosed in a metal housing. External antennas also maximize the internal volume available for other product components. However, they may cost more than internal antennas (especially if RF connectors are needed) and, because they lack the protection of the product housing, they may be prone to mechanical failure due to environmental factors, handling, and so on. The loss of the feedline between the antenna and transceiver must also be considered in the product design.

If an internal antenna is the decision, there are several design choices available. The first choice is often to employ a circuit board trace as an antenna. This is certainly the cheapest approach, because no antenna components need be purchased, handled, or placed on the board, but it is not free, because circuit board area must still be purchased for it. On the negative side, circuit board antennas tend to have significant loss, due to their resistance (caused by the thinness of the copper trace) and their proximity to lossy materials such as the circuit board material itself, nearby ground planes and traces, and other circuit components.

Wire antennas, such as loops and dipoles, allow the antenna to be placed above the plane of the circuit board, improving its efficiency, and they are often a good compromise between the performance of external antennas and the cost of circuit board trace antennas. Wire antennas may require tuning for best performance, due to mechanical shape and size variations that affect their frequency of resonance, and they may need non-metallic physical supports in certain applications.

Specialized antenna components, composed of ceramic materials, are also available. They offer a smaller physical size than that attainable with wire antennas and do not require tuning; however, they are more expensive than wire antennas.

PRODUCT DESIGN FLEXIBILITY

IEEE Std 802.15.4 is designed to have the flexibility to serve in a wide variety of applications—from personal computer peripherals and toys to home automation and industrial controls. These applications, however, may have very different cost, size, and current drain goals, and the ideal design for one may be unsuitable for another. Figure 7–4 is a spider chart showing the relative importance of size, cost, and current drain for three possible IEEE Std 802.15.4 applications. Note that these determinations are examples only and will vary greatly, depending on implementation details and the specific applications chosen within these broad categories.

In this illustrative example, cost is of primary importance in the home automation application, whereas current drain is of the least importance—possibly because the network may have access to mains power. Similarly, current drain is of primary importance in the industrial control application, whereas size is of the least importance—possibly because the application is a large industrial motor controller. Finally, size is the primary consideration in the PC peripheral application—for example, a wireless mouse or joystick—whereas cost is of the least importance. The application designer must make the right trades in these three dimensions to optimize an IEEE Std 802.15.4 product.

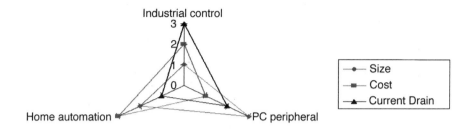

Figure 7–4: Example of relative importance of performance parameters for different applications

Chapter 8   Real-World Issues

...a touch of reality

Users, implementers, and system integrators of IEEE Std 802.15.4 should be aware of some factors that may affect the performance of their systems resulting from interaction with the "real world." Two of these factors are the coexistence of IEEE Std 802.15.4 products with other users of the radio frequency spectrum, and the interfacing of different communication protocols.

COEXISTENCE

Because the bands used by the two IEEE Std 802.15.4 PHY layers are unlicensed in nearly all countries, there are devices employing other communication protocols sharing the bands. In the 868 MHz band, for example, there are proprietary low-data-rate protocols; in the 915 MHz band, proprietary protocols and cordless telephones; and in the 2.4 GHz band, cordless telephones, microwave ovens, WLANs, and other WPANs. In most regulatory regions, devices operating in these unlicensed bands must accept interference to them caused by other services operating in the same band, and their users do not have legal redress for any performance degradation caused by the other services. Although this may be the regulatory reality, it is not the market reality; users purchasing new devices reasonably expect that not only will the new devices operate as desired, but that they also will not disrupt the operation of existing products they may have. A device not meeting these two criteria may have a difficult time remaining on the market. Coexistence is therefore an important economic issue.

Several features of IEEE Std 802.15.4 were designed with coexistence with other services in mind. Although the issue can be considered as two problems (the protection of other services from interference caused by IEEE Std 802.15.4 transmissions, and the protection of IEEE Std 802.15.4 implementations from interference caused by the transmissions of other services), in practice most solutions to one of these problems are also solutions to the other. Some features of IEEE Std 802.15.4 that reduce potential interference to other services are listed in the following paragraphs; much of this list is derived from the Coexistence Annex of the standard.

> *Low Transmitted Power*: Although radio regulations in many parts of the world (including Part 15.247 of the FCC rules [22] allow operation in the 915 MHz

and 2.4 GHz bands with relatively high output power (as much as 1 W in some cases), the typical IEEE Std 802.15.4 device will likely operate with much lower transmit power. A key metric of IEEE Std 802.15.4 is cost, and achieving greater than 10 dBm transmit power in a low-cost system-on-chip (SOC), although feasible, will be economically disadvantageous. Furthermore, FCC and European regulations (ETSI EN 300 220-1 V1.3.1 [5], CEPT Recommendation 70-03 [4], and ETSI SN 300-328 [6] for out-of-band emissions make it difficult to employ powers above 10 dBm without additional, expensive filtering of the transmitted spectrum. These factors will limit the distribution of devices with greater than 10 dBm transmit power to a few specialized applications.

- *Channel Alignment:* A concern in many applications of IEEE Std 802.15.4 that employ the 2.4 GHz PHY is coexistence with WLANs using the popular IEEE Std 802.11, particularly those WLAN installations in which the maximum of three non-overlapping WLAN channels are employed. Figure 8–1 shows the alignment between the North American non-overlapping IEEE Std 802.11b channels and IEEE Std 802.15.4 channels. There are four IEEE Std 802.15.4 channels that fall in the guard bands between (or above) the three IEEE Std 802.11b channels (n =15, 20, 25, 26 for North America, n = 15, 16, 21, 22 for the European non-overlapping WLAN channels). Although the energy in this guard space will not be zero, it will be lower than the energy within the channels, and operating an IEEE Std 802.15.4 network on one of these channels will minimize interference between systems. When performing dynamic channel selection (controlled by the upper layers), either at network initialization or in response to channel impairment, an IEEE Std 802.15.4 device will scan a set of channels specified by the "ChannelList" parameter in the MLME-SCAN.request primitive. For IEEE Std 802.15.4 networks that are installed in areas known to have high IEEE Std 802.11b activity, the ChannelList parameter can be defined as one of the previous sets (either [15, 20, 25, 26] or [15, 16, 21, 22], as appropriate) in order to maximize the coexistence of the networks.

- *Channel Selection:* Prior to network formation, the candidate PAN coordinator performs a scan of all channels in its channel list, to identify IEEE Std 802.15.4 PANs already in existence. If a suitable PAN is identified, at the discretion of the network layer the candidate may choose to join that PAN, rather than create a second PAN. This behavior will minimize the number of PANs existing in a band, thereby reducing potential interference to other services. Should no acceptable PANs be identified, however, the candidate PAN coordinator may choose to start a new PAN. It selects an unoccupied channel for the new PAN from the list of channels in the ChannelList

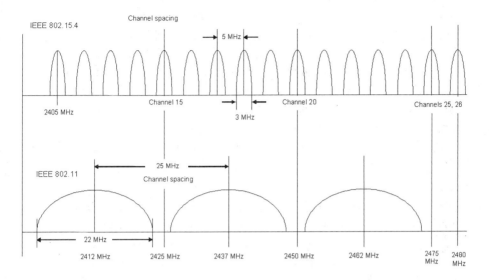

Figure 8–1: Alignment of the North American IEEE Std 802.11 non-overlapping channel plan and IEEE Std 802.15.4 2.4 GHz PHY channels

parameter (which may be determined by an energy detection channel scan), a second behavior that avoids transmission on frequencies occupied by other services.

Should interference appear on the channel occupied by the PAN, the upper layers of the PAN coordinator execute a dynamic channel selection algorithm (not defined in IEEE Std 802.15.4). The PAN coordinator scans the available channels in its channel list and selects a new channel for the PAN. It then returns to the original, impaired channel and makes a broadcast message to the PAN, identifying the new channel to which the PAN is to move. This procedure helps IEEE Std 802.15.4 systems avoid causing interference to, and receiving interference from, other services.

- *Clear Channel Assessment:* The CSMA-CA channel access mechanism in IEEE Std 802.15.4 performs a Clear Channel Assessment (CCA) prior to transmission. The PHY requires that at least one of three CCA methods be used—energy detection (ED) over a certain threshold, detection of a signal with IEEE Std 802.15.4 characteristics, or a combination of the two. Use of the energy detection option improves the coexistence behavior of IEEE Std 802.15.4 with other services by allowing transmission backoff if the channel is occupied by any device, regardless of the communication protocol it may use.

- *Use of Spread Spectrum Techniques:* The direct sequence spreading technique used by IEEE Std 802.15.4 provides some protection for users of narrowband (i.e., less than 200 kHz wide) communication protocols, by spreading the 250 kb/s data transmission over a bandwidth greater than 2 MHz. This reduces the energy of IEEE Std 802.15.4 transmission present in any given narrowband channel, protecting these services. Interestingly, even though its signal is spread, due to its low data rate IEEE Std 802.15.4 transmission bandwidth is comparable to that of IEEE Std 802.15.1 (Bluetooth); as a result an IEEE Std 802.15.4 PAN interferes with a Bluetooth PAN much as a second Bluetooth PAN would—affecting only three of Bluetooth's 79 hops.

- *Link Quality Indication:* IEEE Std 802.15.4 PHY layer standard specifies a Link Quality Indication (LQI). The LQI is performed on each received packet, and it may be implemented using received signal strength, a signal-to-noise ratio estimation, or a combination of these. LQI can be used to detect channel impairment caused by interference on a packet-by-packet basis, providing "real time" information to the upper layers of the channel condition, enabling the device to make an informed decision in dynamic channel selection.

IMPLEMENTATION OF COLOCATED TRANSCEIVERS

An implementer of a system employing two RF transceivers, such as a cell phone incorporating a WPAN, should be aware of interactions that may occur between the two devices. Many applications may require such a design, including those in which the cell phone incorporates a PAN coordinator and is used as the gateway device between the WPAN and the Internet. These interactions can be due to both radiated and conducted effects.

A primary concern when placing two RF transceivers in the same product (or, indeed, in close proximity) is the effect the radiated transmitted noise of one device may have on the receiver of the second device. All transmitters transmit, in addition to the desired signal, some amount of wideband noise, often for many hundreds of megahertz around the desired signal. Ordinarily, because the amplitude of this noise is so low, it is of no consequence, and it is not economically justifiable to insert filters or other circuits to reduce it further. However, if a receiver (or, more specifically, a receiving antenna) is brought very close to the transmitting antenna, the noise can become significant. In this case, the noise may affect the apparent receiving sensitivity of the second device, even though the frequencies of operation of the two transceivers differ greatly. If the communication protocols of the two transceivers are independent, so that the transmission of one device cannot be prohibited while the second device is receiving, special measures, in the form of adding selectivity to the transmitter output, may have to be performed to enable successful operation of the two transceivers.

A second radiated effect is the effect of receiver blocking. Blocking, a strong signal effect (as opposed to a noise effect), occurs when a very strong undesired signal enters a receiver and affects the dc bias of its circuits. This typically causes a loss of gain, leading to a loss of sensitivity. In a sense, this affects the "signal" portion of the receiver detector's signal-to-noise ratio, whereas wideband transmitted noise affects the "noise" portion. Blocking can happen, for example, if a 600-mW cellular transmitter is co-located with a WPAN receiver that is biased for low-power operation. Blocking performance can be improved by adding selectivity before the receiver input, if the offending signal is outside the passband of the victim receiver or, usually less desirable, by increasing the dc bias of the receiver circuits (i.e., increasing power consumption).

A conducted effect that can arise when employing two transceivers results from the sharing of a power supply. In portable devices this is often unavoidable, because the primary supply is a battery and there is rarely room in the size or weight budget for a second battery in the product. However, if one transceiver draws significant current from the shared supply (usually while transmitting), any voltage drop resulting from either internal or external resistance associated with that supply could affect the second transceiver. The textbook example of this phenomenon is the supply voltage modulation caused by a time division multiple access (TDMA) cellular transmitter, such as a global systems for mobile (GSM) transmitter, which, because it is at an audio rate, caused an audible tone in early GSM handsets. This problem can be corrected with suitable supply design, but the addition of a separate voltage regulator for the victim device is often required.

Chapter 9   Concluding Remarks

... summarizing

IEEE Std 802.15.4 was designed to meet the needs of an application space that has received little attention until now—that of wireless communication applications requiring only moderate data throughput, but needing very low power consumption and very low implementation cost. The IEEE Std 802.15.4 Task Group has worked to produce a standard that is at once flexible enough to meet the demands of a wide range of applications—from the wireless sensor networks used in intelligent agriculture and military applications, to the industrial control and monitoring networks, to consumer electronics and home automation uses—while retaining a minimum of costly options and rarely used special features that, while optimizing performance in one application, add complexity to all others. The Task Group believes that by striking the proper balance between specialization and complexity, it has maximized the utility of the standard to the wireless communication industry.

To maximize worldwide utility, IEEE Std 802.15.4 offers two physical layers: the lower PHY, including both the European 868.0–868.6 MHz and North American 902–928 MHz bands, and the upper PHY, in the more widely available 2.4000–2.4835 GHz band. At 868.0–868.6 MHz, a single channel with raw data rates of 20, 100, and 250 kb/s is available; in the 902–928 MHz band, 10 channels, each with raw data rates of 40 and 250 kb/s, are available; and in the 2.4000–2.4835 GHz band, 16 channels, each with a raw data rate of 250 kb/s, are available. Direct Sequence Spread Spectrum (DSSS) is employed in both physical layers to take advantage of the low cost, largely digital implementations that result—and to enable the standard to enjoy future cost reductions due to Moore's Law. In the lower PHY, a conventional, binary BPSK form of DSSS, orthogonally encoded, half-sine-shaped O-QPSK, or root-raised-cosine-shaped Parallel Sequence Spread Spectrum (PSSS) may be employed, whereas in the upper PHY a form of orthogonal coding is employed with half-sine-shaped O-QPSK.

The combination of the relatively low transmitted power expected to be used by most compliant devices, the use of DSSS, the use of CCA techniques prior to network formation, and the low duty cycle of the MAC, should make IEEE Std 802.15.4 devices among the best of neighbors in the coexistence with other wireless technologies.

To achieve the IEEE Std 802.15.4 goals of wide utility without undue complexity, the MAC employs a simple slotted superframe with optional beacons. To support applications requiring guaranteed bandwidth, such as wireless keyboards and mice, time can be reserved in each superframe for specified network devices in guaranteed time slots (GTSs). The superframe time not occupied by beacons or GTSs can be designated as a contention period, in which devices may contend for the channel using a conventional CSMA technique. Contention-based access was chosen over polling-based access to reduce power consumption of the networked devices (by reducing their duty cycle), and to support multiple network topologies with a minimum of complexity.

The standard also specifies security based on the well-respected AES-128 encryption algorithm. Both message encryption and integrity checking are supported; for high-security applications requiring message integrity, the full 128-bit Message Integrity Code (MIC) may be added to each transmitted packet, whereas for applications able to trade some security in exchange for shorter message length the MIC may be truncated to 64 or 32 bits.

IEEE Std 802.15.4 supports both star and peer-to-peer networks, but like all IEEE 802 standards, it does not specify the network layer design. (However, it does specify network association and disassociation mechanisms.) The user may choose to employ a network specified by an industry consortium such as the Zig-Bee Alliance (see "ZigBee" on page 149), or design a specialized network that fits his application requirements.

This combination of worldwide availability, multiple medium access techniques, strong security, and multiple supported network architectures enables a wide variety of applications, and it is the driving force behind the success of IEEE Std 802.15.4. As more and more individuals and organizations discover the value of low-rate wireless personal area networking to improve their productivity and standard of living, this trend will continue.

HIGHER LAYER EXAMPLES

The IEEE 802.15.4 standard is a very flexible, enabling technology supporting different topologies as well as a variety of throughput and latency requirements while enabling low-rate communication among low-power nodes. However, the real application value is derived when this LR-WPAN technology is combined with higher layers to provide a complete solution. As result of the layered approach that commonly followed in the design of communication systems it is possible to combine a base technology such as IEEE 802.15.4 with a variety of higher layer depending on application needs. While the 802.15.4 standard is designed to fit within the 802 family of standards and supports using an 802.2 LLC, some applications may benefit from following different design approaches. Other higher layer communication options that rely on IEEE 802.15.4 compliant PHY and MAC sublayer include ZigBee, Wireless HART, ISA - SP100 and IPv6 over low-power WPAN, to name only a few. These higher layers rely on many of the concepts introduced in this book.

There are many factors that influence implementers to choose a particular high-layer solution over another. These factors are in some cases technology related but more often are market and business dependent. Whatever the factors for choosing a certain low-rate wireless technology are, it is often more advantageous for an implementer to choose a standard solution over a proprietary technology. Standardization helps companies focusing on their core strength while leveraging developments of other companies working together to offer an application specific system solution. Standard technologies allow OEMs to reduce the time to market and development cost by leveraging existing implementations, while often being able to choose among competing solutions.

ZigBee

The ZigBee Alliance combines a large number of companies to create low-cost, low-power, wirelessly enabled monitoring and control products across residential consumer, commercial, and industrial market spaces [29], [30]. Members of the alliance include original equipment manufacturers, chip manufacturers, software developers, solution providers, and independent test houses. All of these companies work together towards developing and promoting worldwide specifications for small interoperable wirelessly enabled devices. The scope of the ZigBee Alliance encompasses not only specifying interoperable higher layer protocols, but also defining certification processes, compliance test specification, and application profile specifications, as well as efforts to market this technology.

The ZigBee Alliance offers a framework that provides the separation of concerns and is built on top of the IEEE 802.15.4 physical layer and MAC sublayer. This framework consists of a network layer, an application support sublayer, an application framework as well as security services (see Figure 9–1), following the simplified five-layer version ISO model shown in chapter 1. The stack supports the simple star topology and also enables hierarchical tree and mesh routing, utilizing many of the features provided by IEEE Std 802.15.4. To reduce implementation size and complexity, the security services are shared between the layers of the ZigBee stack. On the application level, the ZigBee alliance develops interoperable profile specifications for a variety of application areas. These profiles allow devices from different manufacturers to not only coexist but to interoperate. This

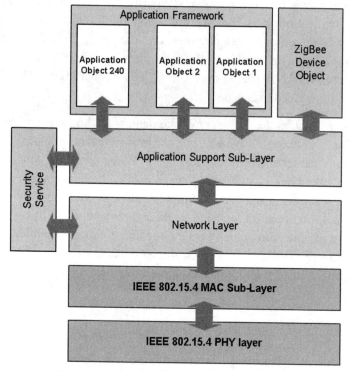

Figure 9–1: ZigBee framework model

combination of providing a flexible higher layer communication stack and application profiles is what enables ZigBee to provide a total solution and therefore allow users of this technology (the OEMs) to focus on their core strength, the product's specific application.

Wireless HART

Wireless HART is a specification being developed within the HART® Communication Foundation (HCF). Wireless HART is the wireless version of this widely used wired protocol in industrial automation. The HART Communication protocol is used today in over 20 million installed process automation devices worldwide. It is expected that the addition of the wireless capability will enable a new the level of connectivity for industrial monitoring and control applications.

The market segments that the Wireless HART technology targets includes a wide range of industries with focus in process automation such as petrochemical, chemical, oil and gas, refining, pulp and paper, pharmaceutical, and so on. Due to the nature of the applications involved, the wireless HART initiative requires the use of interoperable, robust, reliable, and secure technology to achieve its goals. With these requirements in mind, the Wireless HART standardization committee decided to use the IEEE Std 802.15.4 2.4 GHz PHY and MAC. In order to further improve communications reliability and mitigate coexistence issues for the industrial environment, the network layer used in Wireless HART implements a channel hopping methodology. This mechanism allows the regular mesh forming peer-to-peer links to communicate through the possible 16 available DSSS channels. In addition, the MAC is extended to use a version of time division multiplexing suitable for the throughput and latency requirements needed in the HART protocol [31].

ISA—SP100

In February 2005, the Instrumentation, Systems, and Automation Society (ISA) created a committee called the SP100 / Wireless Systems for Automation chartered with the goal of creating a standard for industrial wireless monitoring and control. The application space for the industrial world is summarized in Figure 9–2 [32].

The goal of this committee is to generate a set of wireless communication standards for the industrial world that includes not only process automation, but also factory automation with scope ranging from condition-based monitoring to

Category	Class	Application	Description
Safety	0	Emergency Action	Always critical to plant operation
Control	1	Closed loop regulatory control	Often critical
Control	2	Closed loop supervisory control	Usually noncritical
Control	3	Open loop control	Human in the loop
Monitoring	4	Alerting	Short-term consequence
Monitoring	5	Monitoring	No immediate consequence

Figure 9–2: SP100 application space

real-time control. Typical market segments for these applications includes oil and gas, chemical, manufacturing, automotive, pharmaceutical, food processing, aerospace, and so on.

At the time of finishing the second edition of this book, the standard committee has not settled in the physical layer that will be specified. Out of the 25 preliminary proposals submitted in the third quarter of 2006, 18 were supporting the use of IEEE Std 802.15.4 2.4 GHz PHY and MAC (with some MAC extensions to allow the formation of controlled network topologies that supports 10 ms latencies).

IETF IPv6—LoWPAN

In late 2004, the internet engineering task force created a new working group with the goal of defining a specification for a communications protocol that utilizes transmission of IPv6 packets over IEEE Std 802.15.4 WPAN [33]. The motivation of this effort started from commercial markets in need of wireless connectivity based on the ubiquitous Internet Protocol. Some of the features of this specification include:

- IP adaptation/Packet Formats and interoperability
- Addressing schemes and address management
- Routing in dynamically adaptive topologies
- Discovery (of devices, of services, etc.)

References

[1] Braley, Richard C., Gifford, Ian C., and Heile, Robert F., "Wireless Personal Area Networks: An overview of the IEEE P802.15 working group," *ACM Mobile Computing and Communications Review*, vol. 4, no. 1, January 2000, pp. 26–34.

[2] Callaway, Ed, Bahl, Venkat, Gorday, Paul, Gutiérrez, José A., Hester, Lance, Naeve, Marco, and Heile, Robert, "Home Networking with IEEE 802.15.4, a Developing Standard for Low-Rate Wireless Personal Area Networks," *IEEE Communications Magazine*, special issue on Home Networking, vol. 40, no. 8, August 2002, pp. 70–77.

[3] Ennis, G., "Impact of Bluetooth on 802.11 direct sequence," *IEEE P802.11-98/319*, 1998.

[4] European Conference of Postal and Telecommunications Administration (CEPT), European Radiocommunications Committee; Relating to the Use of Short Range Devices (SRD). CEPT/ERC/REC Recommendation 70–03. December 2002.

[5] European Telecommunication Standards Institute, Electromagnetic compatibility and Radio spectrum Matters (ERM); Short Range Devices (SRD); Radio equipment to be used in the 25 MHz to 1 000 MHz frequency range with power levels ranging up to 500 mW; Part 1: Technical characteristics and test methods. ETSI EN 300 220-1 V1.3.1. Sophia-Antipolis, France: European Telecommunication Standards Institute, September 2001.

[6] European Telecommunication Standards Institute, Electromagnetic Compatibility and Radio Spectrum Matters (ERM); Wideband Transmission Systems; Data Transmission Equipment Operating in the 2,4 GHz ISM Band and Using Spread Spectrum Modulation Techniques. ETSI EN 300-328. Sophia-Antipolis, France: European Telecommunication Standards Institute. 2001.

[7] Gutiérrez, José, Naeve, Marco, Callaway, Ed, Bourgeois, Monique, Mitter, Vinay, and Heile, Robert F., "IEEE 802.15.4—A Developing Standard for Low-Power, Low-Cost Wireless Personal Area Networks," *IEEE Network Magazine*, vol. 15 no. 5, September/October 2001, pp. 12–19.

[8] Heegard, Chris, Coffey, John (Seán) T., Gummadi, Srikanth, Murphy, Peter A., Provencio, Ron, Rossin, Eric J., Schrum, Sid, and Shoemake, Matthew B., "High-performance wireless ethernet," *IEEE Communications*, vol. 39, no. 11, November 2001, pp. 64–73.

[9] Howitt, I., "Bluetooth performance in the presence of 802.11b WLAN," *IEEE Transactions on Vehicular Technology*, vol. 51, 2002.

[10] Howitt, I., "WLAN and WPAN Coexistence in UL Band," *IEEE Transactions on Vehicular Technology*, vol. 50, 2001, pp. 1114-1124.

[11] Howitt, I., and Gutiérrez, J. A., "IEEE 802.15.4 low rate wireless personal area network coexistence issues," *Proceedings of the WCNC 2003*, 2003.

[12] Howitt, I., Mitter, V., and Gutiérrez, J., "Empirical study for IEEE 802.11 and Bluetooth interoperability," *IEEE Spring VTC 2001*, Rhodes, 2001.

[13] IEEE Std 802.11™-1999, IEEE Standard for Information technology—Telecommunications and information exchange between systems—Local and metropolitan area networks—Specific requirements—Part 11: Wireless LAN Medium Access Control (MAC) and Physical Layer (PHY) specifications.

[14] IEEE Std 802.15.1™-2002, IEEE Standard for Information technology—Telecommunications and information exchange between systems— Local and metropolitan area networks—Specific requirements—Part 15.1: Wireless Medium Access Control (MAC) and Physical Layer (PHY) Specifications for Wireless Personal Area Networks (WPANs).

[15] IEEE Std 802.15.4™-2003, IEEE Standard for Information technology—Telecommunications and information exchange between systems—Local and metropolitan area networks—Specific requirements—Part 15.4: Wireless Medium Access Control (MAC) and Physical Layer (PHY) Specifications for Low-Rate Wireless Personal Area Networks (WPANs).

[16] Karaouz, Jeyhan, "High-rate Wireless Personal Area Networks," *IEEE Communications*, vol. 39, no. 12, December 2001, pp. 96–102.

[17] Marks, Roger B., Gifford, Ian C., and O'Hara, Bob, "Standards in IEEE 802 unleash the wireless internet," *IEEE Microwave*, vol. 2, no. 2, June 2001, pp. 46–56.

[18] O'Hara, Bob, Petrick, Al, *IEEE 802.11 Handbook: A Designer's Companion.* New York: Standards Information Network, IEEE Press, 1999.

[19] Poor, Robert D., *Embedded networks: Pervasive, low-power, wireless connectivity*, Ph.D dissertation, 2001, Massachusetts Institute of Technology, Cambridge, MA.

[20] Siep, Tom, *An IEEE Guide: How to Find What You Need in the Bluetooth™ Spec.* New York: Standards Information Network, IEEE Press, 2000.

[21] Siwiak, Kai, *Radiowave Propagation and Antennas for Personal Communications*, 2nd ed., Boston: Artech House.1998.

[22] U.S. Code of Federal Regulations, vol. 47, sec. 15.247. Washington, D.C.: U.S. Government Printing Office. 2001.

[23] U.S. Code of Federal Regulations, vol. 47, sec. 15.249. Washington, D.C.: U.S. Government Printing Office. 2001.

[24] U.S. Department of Commerce, National Institute of Standards and Technology, Specification for the Advanced Encryption Standard (AES). Federal Information Processing Standards Publication 197. November 26, 2001. http://csrc.nist.gov/publications/fips/fips197/fips-197.pdf.

[25] van Nee, Richard, "A new OFDM standard for high rate wireless LAN in the 5 GHz band," *Proc. IEEE Veh. Tech. Conf.*, 1999, vol. 1, pp. 258–262.

[26] Zimmermann, Hubert, "OSI reference model—The ISO model of architecture for Open Systems Interconnection," *IEEE Trans.Commun.*, vol. COM-28, no. 4, April 1980, pp. 425–432.

[27] Zyren, J., "Reliability of IEEE 802.11 Hi rate DSSS WLANs in a high density Bluetooth environment," *Bluetooth'99*, 1999.

[28] Schwetlick, Horst and Wolf, Andreas, "PSSS—Parallel Sequence Spread Spectrum Application in RF communication," Proceedings from the *International Symposium on Signals, Systems, and Electronics (ISSSE)* 2004.

[29] Craig, William C., "ZigBee: Wireless Control That Simply Works," ZigBee Alliance white paper.

[30] http://www.zigbee.org

[31] http://www.hartcomm.org

[32] http://www.isa.org

[33] http://www.ietf.org/html.charters/6lowpan-charter.html

Glossary

access control list: A table used by a device to determine which devices are authorized to perform a specific function.

ad hoc network: A wireless network composed of communicating devices without preexisting infrastructure. An ad hoc network is typically created in a spontaneous manner and is self-organizing and self-maintaining.

advanced encryption standard: The symmetric encryption algorithm specified by the U.S. government in Federal Information Processing Standard (FIPS) 197. (http://csrc.nist.gov/encryption/aes/)

application layer: The top layer in the open system interconnection (OSI) data communication model, in which application functions are performed that directly affect the end user.

association: The service used to establish a device's membership in a wireless personal area network.

authentication: The service used to establish the identity of one device as a member of the set of devices authorized to communicate securely to other devices in the set.

authentic data: Data with a source verifiable through cryptographic means.

bandwidth: The difference between the highest and lowest frequencies employed in a communication channel. More informally, the largest amount of data that can be sent per unit time over a communication channel.

baseband: A communication channel through which constant-potential symbols (e.g., dc voltages) are sent, without frequency shifting (e.g., modulation of a high-frequency carrier); the band of frequencies occupied by the signal before it modulates the carrier.

baud rate: A measure of signaling speed equal to the number of discrete conditions or signal events per second; the number of symbols per second in a communication system.

bit rate: A measure of the amount of information per second in a communication system. Also, the number of binary symbols per second in a communication system.

Bluetooth: A Wireless Personal Area Network specification sponsored by the Bluetooth Special Interest Group (http://www.bluetooth.org) and standardized as IEEE Std 802.15.1.

broadband: Having a large bandwidth when compared to a reference; especially, a modulated signal having a large bandwidth when compared to its associated baseband signal. More informally, a high bit rate suitable for multimedia applications.

clear channel assessment: An evaluation of the communication channel prior to a transmission to determine if the channel is occupied.

confidentiality: Assurance that communicated data remains private to the parties for whom the data are intended.

coordinator: A full function device (FFD) with network device functionality that is also capable of providing synchronization services through the transmission of beacons and the creation of a superframe structure. If a coordinator is the principal controller of a PAN, it is called the PAN coordinator.

coverage area: The area where two or more IEEE Std 802.15.4 units can exchange messages with acceptable quality and performance.

cyclic redundancy check (CRC): An error detection code transmitted with a block of data in order to detect corruption of the data.

data integrity: Assurance that received data have not been modified from their original form.

data link layer: The layer between the Physical and Network layers of the open system interconnection (OSI) reference model for communication between computer systems, providing the functional and procedural means to establish, maintain, and release data links between network entities.

disassociation: The service that removes an existing association.

frame: The format of aggregated bits from a MAC sublayer entity that are transmitted together in time.

full function device (FFD): A device capable of operating as a coordinator or network device, implementing the complete protocol set.

IEEE: *See:* **Institute of Electrical and Electronic Engineers**

IEEE Std 802.11: IEEE standard that specifies medium access and physical-layer specification for wireless connectivity among fixed, portable, and moving devices within a local area at data rates of 1 and 2 Mb/s.

IEEE Std 802.11a: Extension of the IEEE Std 802.11 using orthogonal frequency-division multiplexing (OFDM) modulation, operating in the 5GHz frequency band at data rates of up to 54 Mb/s.

IEEE Std 802.11b: Extension of the IEEE Std 802.11 standard using Direct Sequence Spread Spectrum modulation in the 2.4 GHz frequency band at data rates of up to 11 Mb/s.

industrial, scientific, and medical (ISM) bands: Radio-frequency bands reserved for these applications by international agreement. The ISM bands relevant to WPANs are located at 900 MHz, 2.4 GHz and 5.7 GHz.

Institute of Electrical and Electronic Engineers (IEEE): A non-profit, technical professional association with the goal of advancing global prosperity by fostering technological innovation, enabling members' careers, and promoting community world-wide. The IEEE promotes the engineering process of creating, developing, integrating, sharing, and applying knowledge about electro and information technologies and sciences for the benefit of humanity and the profession (http://www.ieee.org).

integrity code: A data string generated using a symmetric key that is typically appended to data in order to provide data integrity and source authentication (also called a message integrity code).

International Organization for Standardization (ISO): A worldwide, non-governmental federation of national standards bodies, promoting the development of standardization and related activities in the world with a view to facilitating the international exchange of goods and services, and to developing cooperation in the spheres of intellectual, scientific, technological and economic activity (http://www.iso.org).

International Telecommunications Union (ITU): An international organization within the United Nations system in which governments and the private sector coordinate global telecom networks and services (http://www.itu.int).

ISO: *See:* **International Organization for Standardization**

ITU: *See:* **International Telecommunications Union**

key establishment: A process by which two entities securely establish a symmetric key that is known only by the participating entities.

key management: Methods for controlling keying material throughout its life cycle, including creation, distribution, and destruction.

key transport: A process by which an entity sends a key to another entity.

ISM bands: *See:* **industrial, scientific, and medical bands**

logical channel: One of a number of distinct channels on a physical communication link.

Logical Link Control sublayer: The upper portion of the Data Link Layer in the OSI reference model, responsible for the organization of data flow.

low-rate wireless personal area network (LR-WPAN): A wireless personal area network, optimized for low-data-rate applications emphasizing long battery life and low implementation cost.

MAC sublayer: *See:* **Medium Access Control sublayer**

MAC protocol data unit (MPDU): The unit of data exchanged between two Medium Access Control entities.

Medium Access Control sublayer (MAC sublayer): The lower portion of the Data Link Layer in the open system interconnection (OSI) reference model, responsible for acquiring the right to use the underlying physical communication medium.

message integrity code: *See:* **integrity code**

mobile device: A device that uses network communications while in motion.

narrowband: Having a small bandwidth when compared to a reference, especially, a modulated signal having a bandwidth comparable to, or less than, its associated baseband signal. More informally, a low bit rate suitable for voice and data applications.

network device: A reduced function device (RFD) or full function device (FFD) implementation containing an IEEE Std 802.15.4 medium access control and physical interface to the wireless medium.

network layer: The layer between the Data Link and Transport layers of the open system interconnection (OSI) reference model for communication between computer systems, providing routing and switching functions and procedures.

network topology: The logical structure of a network.

Open System Interconnection (OSI): The seven-layer ISO model of architecture for communication between (possibly heterogeneous) computer systems.

orphaned device: A device that has lost contact to its associated personal area network coordinator.

packet: The format of aggregated bits that are transmitted together in time across the physical medium.

PAN coordinator: A coordinator that is the principal controller of a personal area network (PAN). An IEEE Std 802.15.4 network has exactly one PAN coordinator.

parent: The PAN coordinator or coordinator through which a network device has associated with.

payload data: The contents of a data message that is being transmitted.

payload protection: The generic term for the provision of security services on payload data, including confidentiality, data integrity, and authentication.

peer-to-peer network: A substantially homogeneous network in which devices communicate as equals, often employing distributed, multihop routing protocols.

personal area network (PAN): A wireless network for the personal operating space.

personal operating space: The space about a person or object that is typically about 10 m in all directions and envelops the person or object whether stationary or in motion.

physical layer: The layer below the Data Link Layer of the open system interconnection (OSI) reference model for communication between computer systems, providing the physical connections between data link entities.

portable device: A device that may be moved from location to location, but only uses network communications while at a fixed location.

protocol data unit (PDU): The unit of data exchanged between two peer entities.

pseudo-random number generation: The process of generating a deterministic sequence of bits from a given seed that has the statistical properties of a random sequence of bits when the seed is not known.

random number generator: A device that provides a sequence of bits that is unpredictable.

reduced function device (RFD): A device operating with a minimal implementation of the IEEE Std 802.15.4 protocol that is capable only of functioning as a network device.

security suite: A group of security operations designed to provide security services on MAC frames.

service access point: Any entity that provides access to the service interface.

service data unit (SDU): Information that is delivered as a unit through a service access point.

service primitive: The basic unit of service provided between layers of a communication protocol.

star network: A network employing a single, central device through which all communication between devices must pass.

symmetric key: A secret key that is shared between two or more parties that may be used for encryption/decryption or integrity protection/integrity verification depending on its intended use.

transaction: The exchange of related, consecutive frames between two peer medium access control (MAC) entities, required for a successful transmission of a MAC command frame or a data frame.

transport layer: The layer between the Network and Session layers of the open system interconnection (OSI) reference model for communication between computer systems, providing transparent transfer of data between session entities.

wireless local area network (WLAN): A computer communication network spanning at most a campus-sized (~1 km) area, employing radio, infrared, or other wireless physical media.

wireless medium (WM): The medium used to implement the transfer of protocol data units (PDUs) between peer physical layer (PHY) entities of a low-rate wireless personal area network (LR-WPAN).

wireless metropolitan area network: A computer communication network spanning at most a city-sized area, employing radio, infrared, or other wireless physical media.

wireless personal area network (WPAN): A computer communication network spanning the personal operating space, employing radio, infrared, or other wireless physical media.

ZigBee Alliance: An association of companies working together to create a very low-cost, very low power consumption, two-way, wireless communications standard (http://zigbee.org).

Index

Energy 83, 143
Orphan 84, 90
Passive 84, 85
Channel selectivity 59
Chip 44–47, 130, 131
Cipher, symmetric 102
Ciphertext 102
Clear channel assessment (CCA) 37, 62, 63, 143, 147, 158
Cluster 105, 123
Cluster head 31, 68, 119
Cluster-tree 30–31, 34, 35, 68, 105, 119–121, 123
Coexistence 11, 141–144, 147
Coherence 27, 28
Collision 32, 34, 65
Colocated transceivers 138, 144
Command type 99
Association request 85, 99
Association response 99
Beacon request 83, 99
Coordinator realignment 90, 93, 99, 100, 101
Data request 74, 91, 95, 99
Disassociation notification 99
GTS allocation 88, 89
GTS request 88, 99, 101
Orphan notification 90, 91, 99
PAN ID conflict notification 99, 100
Communication range 9, 56–58
Free space maximum 56, 57
Log-normal shadowing 57
Consumer electronics 4, 9, 13, 15, 16, 17, 19, 147
Consumer Electronics Association 16
Contention 62, 88, 111, 148
Contention access period (CAP) 62, 70, 73, 87, 90, 101, 135
Contention-free period 70, 87, 90, 101
Coordinator 158
PAN coordinator 21, 29, 30, 31, 66–76, 83, 86–92, 95, 97, 100, 101, 105–120, 135, 142, 143, 144, 158, 161
Cost considerations, architectural features
Data rates 29
DSSS 27, 28
Duty cycle 25, 26

Error control 128
Low battery cost 25
Low-power transmitter 29
Modulation 26, 27
Protocol 132
Relaxed sensitivity 29
CSMA-CA 32, 65, 70, 73, 88, 100, 143, 148
Cyclic redundancy check (CRC) 33, 96, 158

D

Data frame 33, 35, 75, 97, 98, 110, 162
Data link layer 6, 65, 158, 160, 161
Data rates 5, 39, 40
BPSK low rate 41, 44, 47, 56, 147
Cost considerations 29, 40
Frequency bands 39, 41
High rate 29
O-QPSK 41, 147
Data transmission 78, 144
Despreading 27
Device synchronization 67, 70, 90, 91, 92, 94, 101
DeviceNet 15
Devices
Full Function (FFD) 30, 34, 67, 68, 93, 97, 100, 105–111
Network coordinator 30, 93, 158
Network devices 30, 160
Reduced Function (RFD) 30
Network devices 30, 111, 160
Direct Sequence Spread Spectrum (DSSS) 55, 147
802.15.4 implementation 39
Benefits 27, 28, 39, 127
Binary Phase Shift Keyed (BPSK) 47
Binary Phase Shift Keying (BPSK) 147
Coexistence with other services 28
Cost considerations 28, 39
Despreading 27
Multipath interference 127
O-QPSK 147
Spreading 27, 28, 47, 58, 127
Disassociate 99
Disassociation 65, 85, 86, 148, 158
Distortion
Adjacent and alternate channels 59

Untethered 9, 13, 25

V

Virtual wire 13, 19

W

Wideband FM 27
Wireless hub 13, 15, 21
Wireless sensor 8, 13, 14–15, 17, 20

Wireless Sensor Network 6–12, 114, 118, 147
Wireless Sensor Networks 123
WLAN 4, 5, 40, 65, 127, 141, 142, 162
WPAN 4, 13, 40, 59, 65, 127, 141, 144, 145, 159, 162

Z

ZigBee Alliance 6, 7, 16, 148, 162